Dark Matter - Recent Observations and Theoretical Advances

Edited by Michael L. Smith

Published in London, United Kingdom

IntechOpen

Supporting open minds since 2005

Dark Matter - Recent Observations and Theoretical Advances
http://dx.doi.org/10.5772/intechopen.95645
Edited by Michael L. Smith

Contributors
Alexander Bogomolov, Eugene Terry Tatum, Aloke Kumar Sinha, Peter D. Morley, Michael L. Smith, Firmin Oliveira, Arthur N. James, Georgy I. Burde

First published in London, United Kingdom, 2022 by IntechOpen
IntechOpen is the global imprint of INTECHOPEN LIMITED, registered in England and Wales, registration number: 11086078, 5 Princes Gate Court, London, SW7 2QJ, United Kingdom
Printed in Croatia

British Library Cataloguing-in-Publication Data
A catalogue record for this book is available from the British Library

Additional hard and PDF copies can be obtained from orders@intechopen.com

Dark Matter - Recent Observations and Theoretical Advances
Edited by Michael L. Smith
p. cm.
Print ISBN 978-1-83962-440-7
Online ISBN 978-1-83962-456-8
eBook (PDF) ISBN 978-1-83962-457-5

For EU product safety concerns:
IN TECH d.o.o., Prolaz Marije Krucifikse Kozulić 3, 51000 Rijeka, Croatia, info@intechopen.com or visit our website at intechopen.com.

Meet the editor

Michael Smith received his Ph.D. from Colorado State University and then studied biophysics and computerized data handling at Johns Hopkins University, Maryland. He has been publishing on the topic of the proper interpretation of supernovae type Ia (SNe Ia) observations since 2006. Several of his articles deal with dark energy, dark matter, and Hubble's constant determinations. Lately, he has been investigating the basis for the current Hubble tension as stemming from improper handling of the SNe Ia data and differing interpretations of the Friedmann (FLRW) model. He has also published new fundamental relations over the relations between matter, energy, spacetime, and gravity. He has published more than fifty articles in peer-reviewed journals, edited two books on cosmology, and produced six short videos for public education.

Contents

Preface

Dark matter is a subject confined by the observational context. It is currently only indirectly observed as the puzzling motions of distant galaxies and the unpredictable rotational velocities of stars in neighboring galaxies. Attempts to observe dark matter as traditional sub-atomic particles always fail, leaving this worldly topic a major observational problem. This is very different from ordinary matter, which can be mutated, split into components, destroyed, and resurrected at will by physicists, given enough energy, time, and money.

Dark matter was first postulated by astronomers in the mid-twentieth century to explain the seemingly non-Newtonian character of galactic dynamics. A discrepancy was observed by Fritz Zwicky for the relative velocities of three spiral galaxies circulating within a distant, small galactic group. To calculate the relative velocities of these galaxies, Zwicky made the reasonable assumption that the galactic masses are proportional to their luminosities. In that case, the galaxies behaved as if they were much more densely packed with matter than one could explain in simple terms. Since that report, it has been found that dark matter of some form is often required to properly explain the paths and velocities of large galaxies within galactic groups. Though an intriguing observation, I don't think Zwicky necessarily meant that a new type of matter was warranted but rather that the observed rotational velocities are much faster than expected and much of the matter in these galaxies is not luminous.

Dark matter remained just a thought, not seriously considered a problem for many years, until the publication of the relative velocities of some bright stars of the Andromeda galaxy. The velocities of very luminous stars, which are outside the dense interior so could be observed without interfering with light, do not follow Newton's gravitational law. According to Kepler and Newton, as supported by observations of our planetary neighbors, objects more distant from a large central object should travel more slowly than those closer to the galactic center. This correlation does not hold for Andromeda though; astronomers found that past a critical distance from the center, all stars circulate at about the same angular velocity – meaning distant stars exhibit greater instantaneous velocities than those closer to the galaxy center. This is a real surprise. Similar observations of star motions in other spiral galaxies confirming this have been interpreted as necessitating the presence of something like dark matter. To bring this mystery closer to home, the Dutch astronomer Oort observed that nearby stars in our Milky Way seem to be traveling too fast when one considers only the masses calculated for our luminous neighbors.

The outstanding problem with dark matter is how to relate these interstellar and intergalactic observations to something on Earth. Answering this question has been the goal of many young particle physicists with dozens of aspiring theoreticians proposing a bevy of suggestions often in the form of new microscopic particle types. Up to now, observations of any new, special particle that only interacts via gravity but not electromagnetically, as required to be dark matter, have not been confirmed.

The only observations supporting a new microscopic particle were made by the DAMA/LIBRA experiment. This sensitive method uses underground detectors of thallium doped-sodium iodide crystals housed in a radiation-dampening enclosure to detect novel events over many months. The investigators observe a seasonal signal variation that can be simply explained by the annual difference in the earth's velocity as we circle the sun while traveling through our galaxy. It was reported since that even these results cannot be duplicated. If these counterclaims are true this means that all concrete, worldly evidence attributing dark matter to microscopic particles has evaporated. There are several chapters in this book that go beyond the now-discredited claims for a particle type to explain dark matter observations.

The chapter by E. T. Tatum proposes a direct explanation for dark matter, which should be given consideration. He thinks it is likely that dark matter is the abundant cold dark hydrogen that remains undetected, therefore unaccounted, because we lack the technical ability for detection. He makes the reasonable claim that the total mass of hydrogen, being the most abundant species in our universe, has been and is still being seriously underestimated. He provides several references to support his proposition.

In our chapter about dark matter in spiral galaxies, we hypothesize that the effects attributed to dark matter can be modelled by assuming the translational decay of omnipresent axions. The gravitational redshift due to the orbital decay of innumerable axions in a galaxy is translated into a gravity-like effect. The hypothesis is checked against the behaviours of highly luminous stars in two galaxies with well-mapped translations and a good fit is reported between the model and data.

A. N. James reminds us that the Friedmann approximation of the Einstein Equation, as the FLRW interpretation, should perhaps not be taken too seriously. The level where one considers the universe as isotropic and homogeneous should not include the distances from here to "closer" supernovae emissions. He presents a new model that mitigates some of the problems with the FLRW assumptions and offers a counter explanation for the supernovae emission data; not requiring dark matter nor dark energy. He also correctly points out that the Ωk term of this FLRW model is overinterpreted as representing spacetime curvature rather than the more conservative, mathematical view that it only presents the remainder of the Ωm term after normalization, a view shared by others.

In his chapter, A. K. Sinha deals with the likely possibility that dark matter really is very dark; much is located in black holes. This suggestion seems reasonable since black holes are a strong "magnet" for matter and LIGO/Virgo observations suggest black holes are more common than previously thought. He asks the question of how black holes containing dark matter will evolve with time and presents a considered hypothesis for the eventual fate of some dark matter.

P. D. Morley presents a mathematical description of non-Keplerian orbits that should be observed when objects are circulating about non-spherical or multiple centers. He posits that the most abundant particles in the universe, neutrinos, are unlikely to be produced isotropically in galaxies. Such situations could be observed as stars (and planets) following non-Keplerian orbits due to the unequal gravitational pull by these abundant particles. He then presents his mathematical solution to this special solution.

G. I. Burde considers the processes accompanying cosmic ray propagation on cosmological scales. He develops the basic concepts of "relativity with a preferred frame" from the basis of particle dynamics, electrodynamics, and general relativity.

A. Bogomolov presents his argument that dark matter and dark energy are as much a problem of information as astronomy and physics. He considers the philosophies underlying these two concepts and concludes by merging philosophy with cosmology. This editor tends to agree with the ideas of Bogomolov since now that microscopic particles are off the menu, our question is, "What shall we do next, where in our universe shall we search?"

Michael L. Smith, Ph.D., M.D.Sc.
Umeå University,
Umeå, Sweden

Chapter 1

Introductory Chapter: Introduction to Dark Matter

Michael L. Smith

1. Introduction

Dark matter is a subject defined by the observational context. It, whatever it may be, is currently something only indirectly observed as the puzzling motions of distant galaxies and the unpredictable rotational velocities of stars in neighboring galaxies. Attempts to observe dark matter as something like traditional particles almost always fail leaving this earthly topic as a major observational problem. This is a very different situation from that of ordinary matter that can be mutated, split into components, destroyed and resurrected at will by physicists, given enough energy, enough time and plenty of money.

Dark matter was first postulated by astronomers in the mid-twentieth century to explain the seemingly non-Newtonian character of intergalactic dynamics. The first, widely broadcast, discrepancy was observed by Fritz Zwicky for the relative velocities of three spiral galaxies circulating within a distant, smallish galactic group. To calculate the relative velocities of these galaxies, Zwicky made the assumption that the galactic masses are proportional to the luminosities. In this case, the galaxies behave as if they are much more densely packed with matter than one can explain in simple terms. Since that report, it seems that dark matter of some form is often required to properly explain the paths and velocities of large galaxies within galactic groups [1]. Though intriguing, I do not think Zwicky necessarily meant that a new type of matter was warranted but only that the observed rotational velocities are much faster than expectation and that much of the matter in these galaxies is not luminous.

Dark matter remained just a thought, not even considered a serious question for many years, until publication of the relative velocities of some bright stars of the Andromeda galaxy [2]. The velocities of very luminous stars, which are outside the dense interior so could be observed without interfering light, do not follow the Newton-Kepler laws as these rotate around the galactic center. According to classical gravity, as supported by observations of our planetary neighbors, objects more distant from a large central object should travel more slowly than those closer to the galactic center. This correlation does not hold for stars of the Andromeda galaxy though; astronomers found that past a critical distance from the center, all stars circulate at the similar angular velocities—meaning distant stars travel faster, in the relative sense, than those closer to the galaxy center. This is a real surprise (and seemingly an insurmountable challenge). This and more recent observations of star motions in other spiral galaxies have been interpreted as necessitating the presence of something like dark matter [3]. To bring this mystery closer to home, the Dutch astronomer Oort observed that nearby stars in our Milky Way seem to be traveling too fast when one considers only the mass calculated for our luminous neighbors.

It was when these latter observations were published that the problem of dark matter woke up some astronomers and physicists. The outstanding, current problem with dark matter is how to relate these interstellar and intergalactic observations to something on earth. Answering this question has been the goal of many young particle physicists with dozens of aspiring theoreticians proposing a bevy of suggestions often in the form of new microscopic particle types. These ideas include many brands of super-symmetry, where the borders of the standard model of particle physics are enlarged to encompass the new particle types. The most important type might (or may not) be WIMPS (Weakly Interacting Massive ParticleS). But other than exhibiting gravitational attraction, predictions of WIMPS properties are nearly impossible to constrain.

Unfortunately, observations of any new, special particle that only interacts *via* gravity but not electromagnetically, as required to be WIMPS, are very rare and have not been confirmed. A few observations supporting a new microscopic particle type have been claimed by the DAMA/NaI experiment of Italy, which have been searching for new microscopic particles for the past two decades [4]. This sensitive method uses underground detectors of Th-doped-NaI crystals housed in a radiation dampening enclosure to detect novel events over many months. In this respect, the detector(s) is designed in a manner similar to neutrino detectors, expecting very rare, low-energy interactions with normal matter. DAMA has spawned related experiments, ANAIS-112 in Spain and COSINE-100 in Korea, in related attempts to observe WIMPS. Every once in a while, DAMA reports successful observations of signals that indicate the presence of some type of WIMPS [5].

It has been pointed out by others that DAMA investigators observe a seasonal signal variation that can be simply explained by the annual difference in the earth's velocity as we circle the sun—loosely analogous to the Michelson-Morley experiment. It has also been reported that these claims of particle collisions with the NaI detector cannot be duplicated. If these serious counterclaims are true, this means that concrete, worldly evidence attributing dark matter to microscopic particles has evaporated [6]. Good-bye WIMPS.

There are other mechanisms whereby the motions of stars and galaxies attributed to dark matter can be explained. The exaggerated gravitational redshifts might be the results of orbital decay of innumerable axions in a galaxy, which is translated into a gravity-like effect. One may check the validity of such models, presuming the presence of axions, against the behaviors of highly luminous stars in galaxies with well-mapped translations and observe if the model(s) properly describe the data.

Another very likely explanation for dark matter, which seems rather obvious, is abundant dark H and He, which remains undetected, therefore unaccounted, residing in intra- and intergalactic space. Because most of this H and He reside in the ground state, both are dark and remain undetected, except by gravity, because we lack the technical ability for detection. In line with this thought is the likely possibility that dark matter really is very dark; much is located in black holes. This suggestion seems very reasonable since black holes are a strong "magnet" for matter, perturbing the spacetime within and between galaxies rather than indicating the presence of dark matter. Perhaps both mechanisms are in play?

Perhaps, we take the Friedmann approximation of the Einstein Equation, as the FLRW interpretation, too seriously and too direct, and this is part of the problem. The level of accuracy with which one considers the distances from here to many supernovae emissions is exaggerated invalidating the FLRW model. A universe that is not isotropic and homogeneous should not be expected to be predictable. Also, the Ω_k term of the FLRW model is often overinterpreted as representing spacetime curvature. The Ω_k term rather only presents the remainder of universe after the Ω_m term is calculated and not really spacetime curvature [5, 7]. That is, we cannot trust our distance estimates or trust the claim of flat spacetime and violation of the Newton-Kepler gravitational laws should rather be attributed to poor distance estimates.

DOI: http://dx.doi.org/10.5772/intechopen.101679

Finally, recent observations of many black holes of much smaller mass than well-known supermassive black holes, by LIGO/VIRGO, may require more explicit math than the Kepler approximation to explain the stellar orbital motions. These observations also indicate that "smaller" black holes are much more abundant than previously thought. Such a situation should be observed as stars (and planets) follow non-Keplerian orbits due to the unequal gravitational pull within galaxies. This should also be observed as more rapidly rotating galaxies within galactic groups than predicted from luminosity measurements.

Author details

Michael L. Smith
Umeå University, Umeå, Sweden

*Address all correspondence to: mlsmith55@gmail.com

References

[1] Zwicky F. Republication of: The redshift of extragalactic nebulae. General Relativity and Gravitation. 2009;**41**:207. DOI: 10.1007/s10714-008-0707-4

[2] Rubin VC, Ford WK. Rotation of the Andromeda Nebula from a spectroscopic survey of emission regions. The Astrophysical Journal. 1970;**159**:379. DOI: 10.1086/150317

[3] Persic M, Salucci P, Stel F. The universal rotation curve of spiral galaxies I. The dark matter connection. Monthly Notices of the Royal Astronomical Society. 1996;**281**:27. DOI: 10.1093/mnras/278.1.27

[4] Daley J. Testing DAMA. 2019. Available from: https://www.symmetrymagazine.org/article/testing-dama

[5] Baum S, Freese K, Kelso C. Dark matter implications of DAMA/LIBRA-phase2 results. Physics Letters B;**789**:262. DOI: 10.1016/j.physletb.2018.12.036

[6] Cho A. Is the end in sight for famous dark matter claim? Science. 2021;**374**:805. DOI: 10.1126/science.acx9611

[7] Oztas AM, Smith ML. Re-evaluation of Ω_k of the normalised Friedmann-Lemaître-Robertson-Walker model: Implications for Hubble constant determinations. New Astronomy. 2021;**88**:101609. DOI: 10.1016/j.newast.2021.101609

Chapter 2

The Case for Cold Hydrogen Dark Matter

Eugene Terry Tatum

Abstract

The novel 'Cold Hydrogen Dark Matter' (CHDM) theory is summarized in this chapter. Special attention is paid to the fact that current technology prevents us from directly observing extremely cold ground state atomic hydrogen when it is of sufficiently low density in deep space locations. A number of very recent observations in support of this theory are summarized, including cosmic dawn constraints on dark matter. The importance of the Wouthuysen-Field effect as a probable mechanism for CMB decoupling of hydrogen at cosmic dawn is also stressed. This mechanism does not require a non-baryonic dark matter intermediary. Several predictions for this theory are made for the coming decade of observations and simulations.

Keywords: dark matter, CHDM, primordial hydrogen, cosmic dawn, Wouthuysen-Field effect, cosmology theory, Milky Way, galactic evolution, interstellar medium, hydrogen snow clouds

1. Introduction and background

This chapter will familiarize the reader with the author's 'Cold Hydrogen Dark Matter' (CHDM) theory and summarize a wealth of recent observational data in support of this theory. A discussion of a likely cosmic and galactic evolution scenario for primordial hydrogen will follow, and the chapter will end with several predictions for the coming decade of observations and simulations.

1.1 The theory

In May of 2019, the author was invited to attend a dark matter workshop as part of the World Science Festival. There he had the opportunity to present his CHDM theory to colleagues, after which the theory was published [1], and then first summarized in a chapter of IntechOpen's book entitled *Cosmology 2020 - The Current State* [2].

The theory holds that a certain deep space interstellar, circumgalactic, and intergalactic species of cold hydrogen can be particularly difficult to detect, much less accurately quantify. Specifically, extremely low density cold atomic hydrogen in its lowest ground state (i.e., ground state H I with anti-parallel electron spin) *cannot* emit light and does not collide sufficiently often with its neighboring atoms to create a luminous cloud readily visible from Earth or telescopic satellites. Furthermore, its deep space location vastly reduces its frequency of photon absorption that could be visible from Earth or satellites.

In light of this theory, it may seem paradoxical that a soft glow of H I line (21-cm) emissions can be readily detected throughout our Milky Way galaxy (MW) by radio astronomy [3, 4]. The ultrafine 21-cm line indicates a spin flip transition of the hydrogen electron in its ground state. The lower energy spin state is one in which the electron spin orientation is anti-parallel to the hydrogen nuclear spin orientation. Not surprisingly, 21-cm H I emissions are readily detectable in relatively dense and dynamic (cold, warm or hot) hydrogen clouds, which are particularly concentrated within the galactic bulge and the spiral arms of the disk. These 21-cm radio waves readily pass through dense galactic clouds which block most visible light. In fact, our knowledge of the number and distribution of the spiral arms of the MW and other galaxies is largely predicated on radio astronomy H I line mapping. However, while some of these H I emissions are undoubtedly coming from cold (i.e., slow-moving) hydrogen, that does not fully meet the author's definition of CHDM, precisely because it is relatively *concentrated* atomic hydrogen and, therefore, observable. To put it simply, not all cold hydrogen is dark. Only the very low density, lowest ground state, H I species near the cosmic microwave background (CMB) radiation temperature is dark.

When we make observations well away from the spiral arms, and along lines roughly perpendicular to the MW disc, it is estimated from spectral analysis that the vacuum density of deep interstellar space averages roughly one atom per cubic centimeter [5–7]. Similar observations in lines-of-sight to distant light sources have resulted in an average intergalactic vacuum density estimate of roughly one atom per cubic meter. According to the theory presented here, it is in these interstellar, circumgalactic and intergalactic deep space regions where the majority of the difficult-to-detect, extremely cold, and lowest density hydrogen (i.e., CHDM) must reside.

2. Observational support for the CHDM theory

2.1 MW observations and halo mass calculations

Posti & Helmi reported using *Gaia* data on globular cluster proper motions to calculate relative masses of dark matter and visible matter within a 20 kpc (65,200 light-years) radius halo sphere centered on the MW [8]. **Figure 1** shows this author's schematic representation of the Posti & Helmi sphere, according to their description.

This diagram is roughly to scale, since the MW disk has a radius of approximately 50,000 light-years and an average thickness of approximately 1,000 light-years. The halo is denoted in black, and the rare stars in the halo outside the disk and bulge are also schematically represented.

Posti & Helmi provided data which allow one to calculate a dark matter-to-visible matter ratio of 2.54 within their 20 kpc radius sphere. Normalizing the visible mass (in the form of stars, gas clouds and dust) of the MW disc and bulge to a current best estimate [9] of 250 billion ʘ (solar masses), allows one to calculate a dark matter mass within the Posti & Helmi sphere of 635 billion ʘ. This number is obtained by multiplying 250 billion ʘ by 2.54.

To compare the CHDM theory with these observations, one can assume that a great majority of the estimated single atoms per cubic centimeter of the 20 kpc halo sphere deep space are cold hydrogen atoms. This conservatively implies an average vacuum matter density of 1.67×10^{-21} kg.m^{-3} within a spherical volume of 9.85×10^{62} m^3. Multiplying these two numbers together gives a total halo vacuum mass of 1.645×10^{42} kg. This amounts to roughly 827 billion ʘ within the halo

Figure 1.
Posti & Helmi 20 kpc halo sphere and MW galaxy.

vacuum *outside* the thin slice representing the visible galactic disk and bulge. This estimate is obtained by dividing 1.645 x 10^{42} kg by the solar mass of 1.989 x 10^{30} kg. Even allowing for only 0.77 such atoms per cubic centimeter of the halo outside the visible galaxy, the Posti & Helmi ratio of 2.54 can be met. This is because 827 billion ʘ multiplied by 0.77 is 636.8 billion ʘ. So, an exceedingly low average halo vacuum density of approximately 0.77–1.0 hydrogen atom per cubic centimeter can *dwarf* the 250 billion ʘ of the visible MW stars, gas clouds, and dust! This makes the CHDM-defined species of atomic hydrogen a serious candidate for the 'missing matter' we are currently referring to as dark matter.

2.2 Cosmic Dawn observations of the redshifted H I line

The early cosmic dawn (reionization epoch) spin temperature of primordial neutral atomic hydrogen can be studied by measuring the redshifted 21-cm H I line at frequencies at or near 78 MHz. These line measurements (specifically '21-cm brightness temperatures') correspond to redshifts in the range of $15 < z < 20$. By performing this analysis in the EDGES study, Bowman, et al. [10] have accumulated a wealth of data on the temperature of cosmic dawn primordial hydrogen. They have determined that cosmic dawn hydrogen was temporarily chilled to the low single digits of the Kelvin scale, reaching a nadir at around $z = 17$ (about 180 million years after the big bang). Thus, primordial hydrogen during early cosmic dawn, from about 100–250 million years after the big bang, was temporarily *decoupled* from its usual equilibration with the CMB radiation. This is what made it *visible*.

Barkana [11] has used the Bowman data to *significantly* constrain the mass m_x of dark matter particles according to a baryon-dark matter (b-DM) particle interaction theory. This theory holds that, for *non-baryonic* dark matter to have interacted with baryonic matter in the early universe, it needed to *first* decouple from the CMB temperature, chilling to the point where it could *then* decouple primordial hydrogen from the CMB temperature.

Barkana's dark matter constraints are nicely graphed in Figure 3, page 9 of his review. This finding was a great disappointment to WIMP theorists, since the dark matter constraints effectively eliminate most, if not all, weakly-interactive massive particles. It *also* rules out baryons, except for atomic hydrogen, molecular hydrogen and perhaps He-3. This is because Barkana's dark matter constraints effectively rule out a dark matter particle much greater than 2–3 GeV, which strongly supports the case for atomic hydrogen!

To comprehend the significance of these dark matter constraints, the reader should scrutinize Figure 3 of [11]. They should pay particular attention to the dark matter particle mass m_x corresponding to a cross-section σ_1 value of 10^{-20} cm^2 and a 21-cm brightness temperature $\log_{10}$ value (in mK) of 2.32. Note that these values correspond to a cold dark matter particle matching neutral atomic hydrogen, with a similar low velocity scattering cross-section and a mass-energy of 0.938 GeV. Furthermore, it should be remembered that the redshifted cosmic dawn 21-cm H I line is the *signature* of atomic hydrogen in its ground state.

2.3 McGaugh's argument for a 'purely baryonic universe'

In March of 2018, around the time of the Bowman, et al. EDGES publication, physicist Stacy McGaugh published a brief note [12] which made a cogent and compelling argument for baryonic dark matter. He noted that the high intensity of Bowman's redshifted H I line was a problem for the standard model of cosmology assumption of a non-baryonic b-DM interaction at cosmic dawn. McGaugh correctly pointed out that current atomic theory would indicate such a signal has the maximum T_{21} intensity when the neutral hydrogen fraction X_{HI} is 1 and the spin temperature T_S is equal to the Kelvin temperature. Although McGaugh did not explicitly state that a primordial atomic hydrogen self-interaction (i.e., decoupled colder atomic hydrogen interacting with CMB-equilibrated atomic hydrogen) was a more reasonable explanation for the EDGES study results, his statement that these observations would be 'expected for a purely baryonic universe' clearly carried this implication. He made the point that a dark matter particle outside of the standard particle model (i.e., a non-baryonic particle) was unnecessary to explain the EDGES study findings.

2.4 A cosmic Dawn H I mechanism (the Wouthuysen-Field effect)

The Bowman, Barkana and McGaugh publications, while opening up the possibility for a light baryon to fit within the current dark matter constraints, did not clearly specify a mechanism by which a baryon could have decoupled primordial hydrogen from the cosmic dawn CMB temperature. However, as detailed in the author's original CHDM publications, those mysterious cold baryons could well have been the first of the primordial hydrogen atoms chilled by the first stars of cosmic dawn.

The first stars of cosmic dawn are believed to have been massive blue stars emitting a great deal of ultraviolet (UV) radiation, including Lyman-alpha (Lyα) waves. Anyone with a knowledge of UV radiation effects on atomic hydrogen would have learned of the Wouthuysen-Field effect (WFE) discovered in the 1950s [13].

In short, the author proposes that the WFE is the likely mechanism by which the first primordial hydrogen atoms were decoupled from the CMB during cosmic dawn. A non-baryonic intermediary would not have been required.

In the laboratory, Lyα radiation has the correct energy to trigger, by a multi-step process, the otherwise 'forbidden' transition (i.e., parallel to anti-parallel electron spin orientation) in ground state atomic hydrogen. As mentioned in the Introduction, the anti-parallel electron spin state has the lower energy level. This effect would have decoupled H I from the CMB temperature. Absorption and re-emission of Lyα photons effectively causes a redistribution of the balance between the 21-cm hydrogen ground states, such that a higher percentage of H I has its electron in the anti-parallel electron spin orientation with respect to the proton spin orientation. The net effect would have been to chill the Lyα-absorbing H I well below the CMB temperature and to make its redshifted ultrafine 21-cm line highly visible (whereas, CMB-equilibrated H I is *invisible* [14]).

Thus, the author's CHDM theory incorporates the WFE mechanism as a very reasonable alternative to the standard ΛCDM cosmology theory that *non-baryonic* dark matter chilled first during the dark cosmic epoch (which immediately preceded cosmic dawn) and then interacted with (i.e., chilled) primordial H I during cosmic dawn. The author's proposed WFE mechanism is a more simple and direct explanation. More importantly, it better explains the measured intensity and timing of the redshifted 21-cm signal at the beginning of cosmic dawn. The author proposes that it was the first *stars* which triggered the chilling of CMB-equilibrated H I, rather than exotic non-baryonic dark matter particles. To put it simply, it appears likely that *the 'cold dark matter' interacting with cosmic dawn CMB-equilibrated H I was, in fact, the first of the atomic hydrogen to be chilled by the Lyα radiation of the first stars.*

2.5 The hydrogen snow cloud model

For readers to become comfortable with the idea that cold hydrogen (molecular as well as atomic) could still be invisible and, therefore, missing from the baryonic budget, a brief description of Walker and Wardle's hydrogen snow cloud model [15] is in order. In their model, molecular hydrogen intermixed with small amounts of atomic hydrogen and helium can theoretically condense to a cold, high-density regime where solid or liquid hydrogen can form. By a complex mechanism, an inverted entropy gradient forms, allowing for the cloud outer layers to maintain an equilibrium state such that they are *below* the CMB temperature. Crucially, Walker and Wardle calculate specific model luminosities to be so low that such hydrogen snow clouds are, effectively, 'a type of baryonic dark matter' (their words). While the supercooled outer envelope of these snow clouds is mostly in the form of molecular hydrogen, Walker and Wardle presume there to be less cold H I in the snow cloud interiors. Therefore, effectively, the atomic and molecular hydrogen in these snow clouds is invisible to direct observation (i.e., dark). Moreover, there is, as of yet, no good reason why such hydrogen snow clouds could not be in large quantity within and around galaxies. This is CHDM in a somewhat different configuration, but dark hydrogen nonetheless.

2.6 The new galactic pin scintillation method for observing otherwise dark baryonic matter

In January of 2021, Wang et al. [16] reported first results of an ingenious method to observe otherwise invisible baryonic matter in the MW. By using the Australian Square Kilometer Array Pathfinder (ASKAP), they were able to track radio scintillations of such matter by using distant galaxies as 'locator pins' to map their extent. A degrees-long filamentous cloud was observed. As described by Wang, 'This gas is undetectable using conventional methods, as it *emits no visible light of its own and is just too cold for detection via (usual) radio astronomy.*' Artem Tuntsov, another of the authors, suggested the possibility that their cloud could be a hydrogen snow cloud having undergone tidal disruption by a nearby star [17]. This is a new method which will undoubtedly be used to identify a great deal more MW hydrogen dark matter than has been possible to measure otherwise.

3. Discussion

Given the CHDM theory presented here, it is useful to consider a reasonable cosmic evolution scenario following the CMB emission epoch. We begin with the CMB anisotropy pattern created when the universe was transitioning from a 3,000 K plasma ball to a hot neutral gas of mostly atomic hydrogen.

The denser regions within the CMB anisotropy pattern presumably followed the positive feedback of gravity to become the galaxies, galaxy clusters, and filaments we observe. Meanwhile, the fractal distribution of less dense regions of the CMB anisotropy pattern intricately interlaced with the denser regions presumably became much less dense, owing to ongoing universal expansion and cooling. These cooler primordial hydrogen regions ultimately became the interstellar, circumgalactic, and intergalactic deep space we see today. To summarize, the denser regions of the CMB anisotropy pattern became warm to hot gas clouds and hot stars, while primordial hydrogen atoms of deep space became increasingly cold, following the descending CMB temperature.

During the early cosmic dawn this otherwise *invisible* atomic hydrogen became highly visible while it was decoupled from the CMB temperature by the Lyα UV radiation of the first stars. However, continuing cosmic expansion and cooling eventually brought the increasingly-separated primordial hydrogen atoms back into CMB equilibrium. Therefore, it appears likely that this mostly invisible primordial hydrogen is still present in deep space in great abundance. Its low kinetic energy, low average vacuum density, tiny scattering cross-section, and remoteness from light sources makes it the most difficult chemical species in the universe for us to locate and measure. One can certainly understand why, to this point, *indirect* observations, such as galactic rotation curves, have been necessary to provide support for the existence of dark matter.

A discussion of the cosmic evolution of primordial hydrogen would be incomplete without a specific focus on its role in galactic evolution and dynamics. By the end of the cosmic dark age, the universe, according to the widely-accepted big bang theory, was at a CMB radiation temperature of a little less than 100 K. The first stars of cosmic dawn had not yet ignited. In their place were rapidly-accumulating concentrations of warming primordial hydrogen interspersed with lesser amounts of primordial helium and lithium. Intimately interlaced with these gravitating and warming hydrogen clouds and 'pre-stars' were zones of hydrogen still equilibrated with the CMB temperature and destined to be, with further cosmic expansion, the cold atomic hydrogen of interstellar, circumgalactic, and intergalactic deep space. According to the CHDM theory, this was the slowly-cooling primordial hydrogen destined to be what we are currently calling 'cold dark matter.'

Not surprisingly, because deep space is interlaced between today's stars, galaxies, galaxy clusters and filaments, deep space primordial hydrogen undoubtedly has a gradient of concentrations and temperatures (i.e., velocities). This depends upon its precise interstellar, circumgalactic or intergalactic location. Thus, the deep space interstellar and circumgalactic vacuum density averages approximately one atom per cubic centimeter, and the deep space intergalactic vacuum density averages approximately one atom per cubic meter. Presumably, in the vicinities of particularly dynamic and concentrated star formation, the interstellar primordial hydrogen is warmed up by energetic photons and concentrated stellar winds, to the point where it can become either visible or highly depleted within these zones. This mechanism may account for observations of a 'cored' (i.e., relatively depleted) dark matter, as hydrogen, distribution near the centers of active star-forming galaxies.

The study by Read et al. [18] of nearby dwarf galaxies showed that those which stopped forming stars more than 6 billion years ago tended to be cuspier than those with more recent 'bursty' star activity. The more recently-active galaxy centers tended to show more pronounced dark matter coring. The Read findings agree well with models where dark matter (baryonic or non-baryonic) is heated up and/or redistributed by concentrated active star formation. This interpretation fits nicely with the CHDM theory concerning the galactic evolution of primordial hydrogen. If this is the correct interpretation, then many bizarre properties of non-baryonic dark matter become unnecessary.

With the cosmic and galactic evolution scenario presented above, it should be apparent that the theorized non-baryonic dark matter is not required in order to understand the structural distribution of visible matter we see today. The opposing processes of gravitational clustering and cosmic expansion provided for a divergence of outcomes within the primordial hydrogen. Presumably, approximately one-sixth of this primordial matter gravitationally collapsed sufficiently to become the visible matter portion of our universe, and the remainder of this primordial matter became the CHDM of deep space.

As for predictions having to do with CHDM, the following appear to be likely:

3.1 Improved methodologies for detecting baryonic dark matter

There will be many more creative methodologies invented over the next decade which will indirectly locate and quantify otherwise invisible collections of cold hydrogen in the MW and its circumgalactic medium. Some of these methodologies may even be applicable to nearby galaxies.

3.2 Tightening constraints on dark matter

There will be tightening dark matter constraints around a particle m_x value of 0.938 GeV (i.e., the mass-energy of neutral atomic hydrogen).

3.3 Computer simulations of CHDM

Computer simulations of galaxy formation and evolution which incorporate the CHDM theory presented herein will show excellent correlations with observations, including the coring effect of heating and/or ejecting cold interstellar hydrogen from active galactic centers with bursty star formation.

3.4 No exotic non-baryonic dark matter

No exotic non-baryonic dark matter fitting the observed qualitative and quantitative dark matter constraints will ever be discovered.

4. Summary and conclusions

An explanation has been given as to why very low density, deep space, primordial atomic hydrogen in its lowest energy electron spin ground state should be largely invisible, especially when close to the CMB temperature. Calculations made on the 20 kpc radius Posti & Helmi halo sphere support this theory, especially when one realizes that an average halo vacuum density of only 0.77–1.0 hydrogen atom per cubic centimeter dwarfs the visible matter of the MW galactic disk and bulge.

Further support for the CHDM theory is provided by the EDGES study of the redshifted 21-cm H I line corresponding to cosmic dawn. The cosmic dawn dark matter constraints given in Barkana's review, and McGaugh's arguments for a 'purely baryonic universe,' fit nicely with the theory presented herein. The most reasonable mechanistic explanation for the intensity and timing of the cosmic dawn signal appears to be the Wouthuysen-Field effect. There appears to be no necessity for an exotic non-baryonic intermediary in the creation of this signal.

Exciting recent theoretical work on the hydrogen snow cloud model also provides some rationale for an abundance of extremely cold hydrogen invisible to direct observation. The new galactic pin scintillation method for indirect observation of 'baryonic dark matter' not reliant upon gravitational phenomena is also exciting. These new ideas and methods bode well for additional inventive studies over the ensuing decade.

The discussion section has focused on providing reasonable cosmic and galactic evolution scenarios for primordial atomic hydrogen which clearly do not require a *pre-existing* structural scaffold of non-baryonic dark matter. Primordial hydrogen created the structure we see. A reasonable CHDM explanation for a correlation between 'cored' galactic dark matter distribution and active galactic centers has also been given. This is consistent with observations reported by Read, et al.

Finally, several predictions having to do with CHDM have been made for the coming decade of observations and simulations.

In conclusion, it is worth asking the following question:

'If interstitial cold atomic hydrogen in its lower ground state is qualitatively and quantitatively sufficient to explain dark matter observations to date, do we really need to spend more of our time and money continuing to look for anything else?'

Author details

Eugene Terry Tatum
Independent Researcher, Bowling Green, KY, USA

*Address all correspondence to: ett@twc.com

References

[1] Tatum, E.T. (2019). My C.G.S.I.S.A.H. Theory of Dark Matter. J. Mod. Phys. 10: 881-887. https://doi.org/10.4236/jmp.2019.108058

[2] Tatum, E.T. (2020). Dark Matter as Cold Atomic Hydrogen in its Lower Ground State. In Michael L. Smith (Ed.). *Cosmology 2020: The Current State.* (pp. 93-102). London: IntechOpen. https://dx.doi.org/10.5772/intechopen.91690

[3] Kalberla, P.M.W. and Kerp, J. (2009). The H I Distribution of the Milky Way. Annu. Rev. Astron. Astrophys. 47:27-61. doi:10.1146/annurev-astro-082708-101823

[4] Wikipedia contributors. "Hydrogen line." *Wikipedia, The Free Encyclopedia*. Wikipedia, The Free Encyclopedia, 26 Mar. 2021. Web. 26 Mar. 2021.

[5] Mammana DL. (2000). *Interstellar Space*. Popular Science, New York, p. 220.

[6] Chaisson E and McMillan S. (1993). *Astronomy Today*. New York. Prentice Hall, p. 418.

[7] Pananides NA and Arny T. (1979). *Introduction to Astronomy.* Second Edition. Wiley & Sons, New York, p. 293.

[8] Posti L and Helmi A. (2019). Mass and Shape of the Milky Way's Dark Matter Halo with Globular Clusters from *Gaia* and Hubble. A&A. 621: A56. DOI:10.1051/0004-6361/201833355

[9] Watkins LL, et al. (2019). Evidence for an intermediate-mass Milky Way from *Gaia* DR2 Halo Globular Cluster Motions. The Astrophysical Journal. 873:118-130. DOI:10.3847/1538-4357/ab089f

[10] Bowman JD, et al. (2018). An Absorption Profile Centered at 78 Megahertz in the Sky-Averaged Spectrum. Nature. 555, 67 DOI:10.1038/nature25792

[11] Barkana R. (2018). Possible Interactions Between Baryons and Dark Matter. arXiv:1803.06698v1 [astro-ph.CO]

[12] McGaugh S.S. (2018). Strong Hydrogen Absorption at Cosmic Dawn: The Signature of a Baryonic Universe. arXiv:1803.02365v1 [astro-ph.CO] DOI:10.3847/2515-5172/aab497

[13] Wikipedia contributors. "Wouthuysen–Field coupling." *Wikipedia, The Free Encyclopedia*. Wikipedia, The Free Encyclopedia, 14 Dec. 2020. Web. 26 Mar. 2021.

[14] Baek, S., et al. (2010). Reionization by UV or X-ray sources. A&A. 523: A4. arXiv:1003.0834. DOI:10.1051/0004-6361/201014347.

[15] Walker, M.A. and Wardle, M.J. (2019). Cosmic Snow Clouds: Self-gravitating Gas Spheres Manifesting Hydrogen Condensation. ApJ. 881:69. https://doi.org/10.3847/1538-4357/ab2987

[16] Wang, Y., et al. (2021). ASKAP Observations of Multiple Rapid Scintillators Reveal a Degrees-long Plasma Filament. MNRAS. 502: 3294-3311. https://doi.org/10.1093/mnras/stab139

[17] University of Sidney Press Release. (Feb 04, 2021). Student Astronomer Finds Galactic Missing Matter. https://sciencecodex.com/student-astronomer-finds-galactic-missing-matter-666476

[18] Read, J.I., et al. (2019). Dark matter heats up in dwarf galaxies. Mon. Not. Roy. Astro. Soc. 484: 1401-1420. arXiv:1808.06634v2 [astro-ph.CO] https://doi.org/10.1093/mnras/sty3404

Chapter 3

Dark Matter in Spiral Galaxies as the Gravitational Redshift of Gravitons

Firmin Oliveira and Michael L. Smith

Abstract

Several recent attempts to observe dark matter with characteristics similar to atomic or subatomic particles as Weakly Interacting Massive Particles (WIMPs) have failed to detect anything real over a wide energy range. Likewise, considerations of large, non-emitting objects as the source of most dark matter fall short of expectations. Here we consider the possibility that massless gravitons suffering slow redshift may be responsible for the properties of spiral galaxies attributed to dark matter. Particles such as gravitons will be extremely difficult to directly detect; the best we can envision is measuring this influence on stellar and galactic motions. Since the motions of stars and galaxies are non-relativistic, we can apply our idea to describe the expected large-scale motions using only Newtonian mechanics. Using our assumption about the importance of the graviton, we here describe the well-known Tully-Fisher relationship of spiral galaxies without resorting to hypothesizing exotic WIMPs or invoking modifications of Newtonian dynamics (MoND).

Keywords: dark matter, spiral galaxies, galaxy dynamics, baryonic Tully-fisher relation, Newtonian mechanics, gravitons, BTFR

1. Introduction

The first hint for the existence of dark matter (DM) between galaxies was the observation over eight decades ago from a study of the luminosities and centrifugal velocities of three galaxies, a galactic "triplet" of the Coma Cluster [1]. Fritz Zwicky noted that the motions of these galaxies about a central point could not be properly described using his presumed values for the galaxy matter densities. He, like all astronomers, estimated matter density assuming that the average density is proportional to the observed luminosity. These three galaxies were behaving as if each contained much more matter than calculated from the luminosities. Zwicky proposed that these three galaxies behaved as if under the influence of a type of plentiful, non-radiating matter, hence DM.

Stellar motions in spiral galaxies The existence of significant DM was later required to explain another interesting problem of astronomy—the intragalactic motions of stars within the Andromeda galaxy [2]. For what seems a general rule, the rotational motions of stars about the center of the galactic plane, of spiral-type galaxies, do not follow Kepler's laws of planetary motion. Instead, the more distant stars travel at much greater velocities relative to those of the interior [3]. Distant stars circulate at a more frantic pace than those closer to the center contradicting the planetary laws of Kepler; objects more distant from the center should be traveling slower than those closer. Follow-up observations by Rubin and coworkers of the stellar velocities within many other spiral galaxies confirm this observation as a generality [4]. This is strongly supported by another group in a study of over 1000 galaxies [5]. It still seems uncertain if this grand observation by Rubin also holds for globular and other types of galaxies because observation of individual star motions is more difficult for these galaxy types. Investigators also pointed out that presumed DM explains the rotational velocities of stars in low luminosity galaxies better than for luminous galaxies [5]. The reader should understand that the quandaries of both Zwicky and Rubin depend on the belief that the quantity and distribution of "normal" matter can be correctly estimated from the galactic luminosity.

Proper explanation of these anomalies remains without a satisfactory answer after more than five decades. An answer to this conundrum is considered as utmost importance to both physics and astronomy. Many investigators now call for either new physics or for the presence of considerable DM, of some type, to explain the stellar motions in these galaxies [6]. A solution of Rubin's observations was proposed years ago by utilizing Einstein's General Relativity under the condition of cylindrical geometry [7] but was quickly shown deficient for describing other important properties of spiral galaxies [8]. Another and a more credible explanation of the baryonic-Tully-Fisher relation (BTFR) has been made using Carmeli General Relativity (CGR), which does not require new particle physics nor DM [9].

Dark matter identity If DM is the explanation, the exact nature still defies description after all these years. We divide the proposed solutions into three suspects. The first type would be DM as small atomic or subatomic particles, which are often termed **W**eakly **I**nteracting **M**assive **P**article**s** or (WIMPS) [10, 11]. We include exotic particles, as those predicted by weak-scale supersymmetry, in this group [12]. The terms massive, WIMPS refer to particles with energies significantly above the common atomic and subatomic energies. To be effective DM such particles would obviously have to exhibit measurable mass, be non-relativistic and hence *cold* in nature. They would also have to be dark, that is, not interacting with photons. WIMPS must also exhibit elastic collisions with themselves and with baryons (no clumping allowed), and there would have to be a lot of them—about 4–6 times the total mass of ordinary matter. In addition, because WIMPS are non-interactive, cold and nearly frictionless, a novel explanation is required to explain how the velocities were significantly reduced after the Big Bang. This is a long list of special conditions and novel physics.

Many attempts have been made to detect DM, as WIMPS, in terms of particles at near atomic or subatomic energies. We list some of the latest attempts to detect such a particle here. Years of attempting underground detection of entities from intergalactic origin with sensitive probes have been without any success [13]. Attempted detection of particles resulting from WIMP self-annihilation in our sun also ended without positive results [14, 15]. Detection of signals in large Xe-filled detectors over long periods (time) has also not reported positive results [16, 17]. A recent summary of several programs that have failed to detect WIMPS is available [18]. There is one recent event recorded by the Particle and Astrophysical Xenon TPC (PandaX) collaboration, which might be evidence for WIMPS, which obviously needs to be reproduced [19]. In spite of these several failures, there are proposed programs hopeful to deliver positive results. It is questionable if these searches will be funded.

One proposal will use the planned International Linear Collider, which might be built in Japan, to detect heavy WIMPS up to 3 Tev in size [20]. Another suggestion is to rather concentrate on detecting the lightest neutralino of supersymmetry. According to some people, this hypothesized particle remains the best candidate for dark matter with decent detection prospects [21]. On the other hand, faith in small mass WIMPS around the mass of neutrinos as DM candidates has been ruled out from examination of supernovae observations showing the total neutrino and neutralino mass of the Universe is simply too small to be DM [22]. The best chance for detecting massive WIMPS may come from detecting products from cosmic ray collisions, which are high enough energetically to produce particles with energies from 150 and 500 GeV [23]. But the final argument against WIMPS is the need for special pleading around the existence of WIMPS, which supposedly make up about 80% of matter; about 20% of our Universe. It is *special* that WIMPS are everywhere but not in our solar system, an argument contrary to the Cosmological Principle of Einstein [24]. So we can believe in WIMPS and discard the Cosmological Principle or hold this principle true and discard the notion of WIMPS, but not both.

The second explanation is that matter of the common variety is responsible for the effects of DM, and there is much more common matter than we can detect in spiral galaxies. These dark objects are often referred to as **M**assive **A**strophysical **C**ompact **H**alo **O**bjects (MACHOS). This group includes massive objects such as brown dwarfs, Jupiter-sized planets, neutron stars, and "small" black holes [25]. Some investigators also include white and red dwarfs in this group since those inhabiting other galaxies are invisible to our telescopes but would exert gravitational attraction nonetheless [26]. One should also include large clouds of known atomic particles, such as neutral hydrogen and helium. Large clouds of atoms and small molecules may well be undetected in unlit regions of intra- and intergalactic space. Investigations into the MACHO hypothesis have waned the past few years but might revive if data from two planned satellites support the MACHO scenario. There are two programs may uncover data critical to the MACHO hypothesis. The first is the James Webb satellite, rescheduled (again) for launch later this year. This satellite will be able to detect objects emitting light in the near and mid-infrared range, thus allowing a better assay of the population of dark stars and Jupiter-sized planets [27]. Another satellite with a detection range in the visible and near-infrared regions is the Nancy Grace Roman space telescope. This platform will be able to collect data over a much greater volume than any current instrument, be able to detect new Jupiter sized planets and so should also add to our knowledge of MACHOS and hence DM [28]. The planned launch date is sometime in 2025, so it may be collecting data before the James Webb telescope. Finally, it has been pointed out that the methods previously used to calculate the MACHO density have not been optimal, with better observational methods and more satellite data, this explanation of DM may rebound in popularity [29].

One key relationship associated with the presence of luminous matter in galaxies is the Tully-Fisher relation (TFR), which is something like

$$L \propto V_\theta^{3.5-4} \tag{1}$$

where L is the apparent luminosity of the stars in the galactic plane, and V_θ is the average rotational velocity of these stars. This holds true even when mid-infrared and microwave radiations are assayed [30]. This relationship is sometimes written as

$$M \propto V_\theta^{3.5-4} \tag{2}$$

where M is the estimated mass of luminous stars and modeled gas in that galaxy. This empirical finding is thus a broad correlation between the rotational velocities as determined by the Doppler shift and the baryonic matter content of many spiral galaxies [31]. This empirical correlation is sometimes termed the baryonic-Tully-Fisher relation (BTFR).

The third explanation is gravitons. This particle was first proposed by Feynman to modulate the interactions of gravity in analogy to the photon, which modulates interactions between ordinary matter [32]. A graviton is the agent of the gravitational field having zero rest mass traveling at the speed of light in a vacuum. Since a graviton has never been observed, the bulk properties may be like those of axions or even neutralinos [33]. These are all hypothesized "particles" not exhibiting properties of ordinary or DM other than gravitational attraction [34]. The graviton has also been hypothesized as a type of heavy particle outside the standard model (SM) of physics [35]. In general, not much is known about these particle types placing our culture in a situation similar to our understanding of WIMPS. Unlike the frustrating studies of WIMPS, the study of gravitons has been generally overlooked. Here we suggest a simple description of gravitons in action to better explain the BTFR and hopefully stir up more interest in this line of inquiry.

Finally, there are several people who have expressed their concern that the searches for new particles, WIMPS, and other gravity sources have failed and that future investigations are not likely to do any better [36]. In the meantime, WIMPS and MACHOS are suggested solutions over which many investigators lavish time and imagination, some can even directly profit from broadcasting their ideas [37]. Dark matter and dark energy have made their way into YouTube pop-science culture, which helps support some younger investigators [38]. The topic of dark matter has even been unofficially designated as the calendar day: October 31.

To properly model the Zwicky and Rubin observations, people have proposed theories that do not invoke any particle type but rather modify the fundamental equation of gravity. Some look to modified gravity (MoND) as a revolutionary answer to explain problems on scales larger than our solar system—the galactic scale [39]. Some of the problems with MoND have been addressed for public consumption [40]. This topic, however, is outside the interest of this chapter.

2. Theory

We suggest that gravitons, as agents of the gravitational field, may be modeled as a dense soup of massless particles that nonetheless interact with massive particles. When gravitons travel in a gravitational field, modeled by the equivalence principle as an accelerating system [41], over a short time period $\delta t = \delta r/c$, where the acceleration a at a point r in the field is given by $a = -GM/r^2$, they must experience an energy loss of $\delta\Xi$ due to motion in that field. We express this differentially as

$$\delta\Xi = -(mc^2)\frac{\delta v}{c} = (mc)a\,\delta t = -\frac{GMm}{r^2}\delta r, \tag{3}$$

where m is the total relativistic graviton mass, δv is the change in velocity of the system observed from an inertial reference frame, G is Newton's gravitational constant, M is the baryonic mass of the field source, r is the distance between the center of the source and the location of the moving gravitons, δt is the short travel time of the gravitons at speed c over distance δr. The energy change is a loss (negative) because δv is in the same direction as the motion of the gravitons, so that for an inertial observer moving with velocity δv, the energy of the graviton is redshifted.

Integrating Eq. (3) between radial positions r_0 and r, where $r_0 \leq r$, we obtain the total graviton energy change $\Delta\Xi$, given by

$$\Delta\Xi = \int_{v_1}^{v_2} -\left(mc^2\right)\frac{\delta v}{c} = \int_{r_0}^{r} -mc\left(\frac{GM}{r^2}\right)\frac{\delta r}{c} = -GMm\left(\frac{1}{r_0} - \frac{1}{r}\right). \tag{4}$$

Eq. (4) describes the gravitational redshift of the graviton energies as these travel from a position r_0 of lower (more negative) to a position r of higher (less negative) potential energy consistent with energy conservation. For simplicity, we assume here that the mass M is a point mass.

Consider the energy equation for a galaxy of baryonic mass M interior to position r_0 with a small baryonic mass m in a circular orbit of radius r. The gravitons traversing the distance from the interior mass to the orbiting mass at speed c have a decrease in energy $\Delta\Xi$ given by Eq. (4). The total orbital energy is given by

$$\frac{1}{2}mv^2 - \frac{GMm}{r} + K_g\,\Delta\Xi = \frac{1}{2}mv^2 - \frac{GMm}{r} - K_gGMm\left(\frac{1}{r_0} - \frac{1}{r}\right) = E, \tag{5}$$

where $K_g > 0$ for $r \geq r_0$ and $K_g = 0$ for $r < r_0$. K_g is a galaxy dependent dimensionless coefficient, which is necessary to fit the theory with observation, but the actual physical reason for it is not known at this time.

Multiplying Eq. (5) by $2/m$ and moving all terms except v^2 to the right-hand side with the total orbital energy $E = -GMm/2r$, we obtain

$$v^2 = \frac{GM}{r} + 2K_gGM\left(\frac{1}{r_0} - \frac{1}{r}\right). \tag{6}$$

The properties of a typical spiral galaxy are more easily observed than other types and can be described by the final radial distance r_f, at which the final rotational velocity is v_f and the orbiting mass is m. Substituting v_f for v and r_f for r into Eq. (6) and solving for the viral mass M_{vir}, we obtain

$$M_{vir} = \frac{v_f^2 r_f}{G} = M\left(2K_g\left(\frac{r_f}{r_0} - 1\right) + 1\right), \tag{7}$$

from which we get the apparent dark matter M_d in the galaxy,

$$M_d = M_{vir} - M = 2MK_g\left(\frac{r_f}{r_0} - 1\right). \tag{8}$$

Solving Eq. (7) for r_0, and simplifying yields,

$$r_0 = \frac{2K_g r_f}{\left(\left(v_f^2 r_f\right)/GM\right) + 2K_g - 1}. \tag{9}$$

These results ideally require virtually all the galaxy baryonic mass M to be interior of the radial distance r_0, with only small test masses beyond r_0. Additionally, determining the actual mass interior to r_0 and the velocity v_f is not an exact science. Thus, proving the validity of Eq. (9) for determining r_0 can be difficult.

Here we relate r_0 to the baryonic Tully-Fisher relation (BTFR) [10]. At $r = r_0$, by squaring Eq. (6) and putting it into an inverted form of the BTFR, we get

$$v^4 = MA^{-1} = GM\left(\frac{GM}{r_0^2}\right) = GMa_0, \tag{10}$$

where $A \approx 50 M_{\odot} km^{-4} s^4$ and M in solar mass units [42, 43]. For v in ms^{-1}, M in kg and r_0 in m, $a_0 = GM/r_0^2 = (GA)^{-1} \approx 1.507 \times 10^{-10} ms^{-2}$, where we again emphasize that we are assuming that M is the mass interior to r_0, which is an approximation. However, the BTFR is also an approximation, albeit a very good approximation from a large amount of galaxy rotation data.

For galaxies in general, we model the mass density function $\rho(r)$ as spherically symmetric. The mass $M(r)$ of the galaxy within the radial distance r is given by

$$M(r) = \int_0^r 4\pi\rho(u)u^2 du. \tag{11}$$

With mass density $\rho(r)$, the graviton energy loss Eq. (4) between r_0 and r is expressed by

$$\Delta\Xi(r) = -\int_{r_0}^r mGM(u)\frac{du}{u^2} = -\int_{r_0}^r mG\left(\int_0^u 4\pi\rho(s)s^2 ds\right)\frac{du}{u^2}. \tag{12}$$

Using Eqs. (11) and (12), Eq. (6) becomes

$$v^2 = \frac{GM(r)}{r} - \frac{2K_g\Delta\Xi(r)}{m} = \frac{G}{r}\int_0^r 4\pi\rho(s)s^2 ds + \int_{r_0}^r \frac{2K_g G}{u^2}\left(\int_0^u 4\pi\rho(s)s^2 ds\right)du. \tag{13}$$

To obtain r_0, solve the following equation iteratively,

$$r_0^2 = \frac{G}{a_0}\left(\int_0^{r_0} 4\pi\rho(s)s^2 ds\right). \tag{14}$$

Solving Eq. (13) for K_g yields the equation,

$$K_g = \frac{v_{flat}^2 - GM/r_f}{\int_{r_0}^{r_f}\left(\frac{2G}{u^2}\right)\left(\int_0^u 4\pi\rho(s)s^2 ds\right)du}, \tag{15}$$

where v_{flat} is the velocity of the flat part of the rotation curve.

3. Results

We performed nonlinear modeling of the graviton model, Eqs. (13)–(15), with the velocity rotation curve data for spiral galaxies NGC 3198 and the Milky Way. The mass density function $\rho(r)$ that we use is defined by,

$$\rho(r) = \left(\frac{M}{(4/3)\pi}\right)\sum_{j=1}^n\left(\left(\frac{b_j}{a_j^3}\right) exp\left(-\frac{r}{a_j}\right)\right), \tag{16}$$

where the total baryonic mass equals M over the total radial range of the observational data, a_j are scale lengths, and b_j are weights, where the weights sum to unity. Determinations for the mass $M(r, r_1)$ within the positions r_1 and r, where $0 \le r_1 \le r$, must be normalized, so the mass is computed from Eq. (16) in the form

$$M(r, r_1) = \left(\frac{M}{\int_0^{r_f} 4\pi\rho(u)u^2 du}\right)\int_{r_1}^r 4\pi\rho(s)s^2 ds, \tag{17}$$

where r_f is the largest radial distance of the data: the galaxy *edge*.

NGC 3198 For the analysis of NGC 3198 rotation data, with stellar mass $2.3 \times 10^{10}\,M_\odot$ and mass of gas $0.63 \times 10^{10}\,M_\odot$, we have a total baryonic mass of $M_{N3198} = 2.93 \times 10^{10}\,M_\odot$ [44]. For the rotation curve data of NGC 3198, we use data from the thesis of Begeman [45]. We set $n = 4$, with the scale lengths $a_1 = 1.1$ kpc, $a_2 = 1.27$ kpc, $a_3 = 1.35$ kpc, $a_4 = 1.5$ kpc, and the weights are $b_1 = 0.4$, $b_2 = 0.3$, $b_3 = 0.2$ and $b_4 = 0.1$. By solving Eq. (14) we obtain $r_0 = 4.254$ kpc, the radial distance where the acceleration equals $a_0 = 1.507 \times 10^{-10} ms^{-2}$. Solving Eq. (15), we obtain a value of $K_g = 0.421$. The result using the NGC 3198 data is presented in **Figure 1**. The fit, not necessarily optimal, has a Mean Absolute Error of $MAE = 4.322 kms^{-1}$, expressed by $MAE = (1/n)\Sigma^{n}_{j=1}\|O_j - P_j\|$, where O_j are the observation values and P_j are the model predictor values. The square of the correlation coefficient between the observed v_{obs} and modeled v_{mod} rotational velocities is $r^2 = (corr(v_{obs}, v_{mod}))^2 = 0.986$.

Milky Way Galaxy A fit of the graviton model to the velocity rotation curve data for the Milky Way is shown in **Figure 2**. The mass for our galaxy is approximated by the BTFR given by Eq. (10) with a flat rotational velocity of $v = 202 kms^{-1}$, which yields a mass of $M_{Galaxy} = 8.328 \times 10^{10}\,M_\odot$. The nonlinear curve fit to the rotation curve data [46] is made with $n = 6$, with the scale lengths given by $a_1 = 0.11$ kpc, $a_2 = 0.7$ kpc, $a_3 = 1.71$ kpc, $a_4 = 1.82$ kpc, $a_5 = 1.83$ kpc, and $a_6 = 1.84$ kpc. The weights used are $b_1 = 0.095$, $b_2 = 0.1$, $b_3 = 0.2$, $b_4 = 0.3$, $b_5 = 0.3$, and $b_6 = 0.005$, where the weights sum to unity. Solving Eq. (14), we find the radial distance where the acceleration to be $a_0 = 1.507 \times 10^{-10} ms^{-2}$ is $r_0 = 8.157$ kpc. Then, using this value for r_0, we solve Eq. (15) to obtain a value of $K_g = 0.418$. The fit of the graviton model, not necessarily optimal, has an error of $MAE = 6.33 kms^{-1}$ for the fit up to r_0 and for the entire range of the data up to 1.28×10^3 kpc, the error is

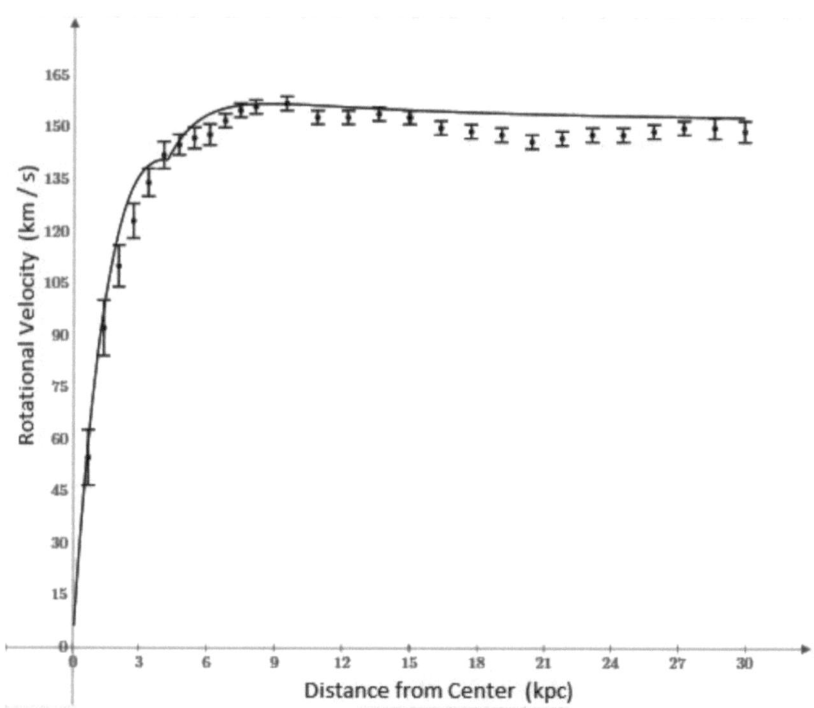

Figure 1.
The velocity vs. radial distance from the galactic center of NGC 3198. The data points and errors are from [45]. The solid curve is a good fit to the data with the graviton model, Eqs. (13–15).

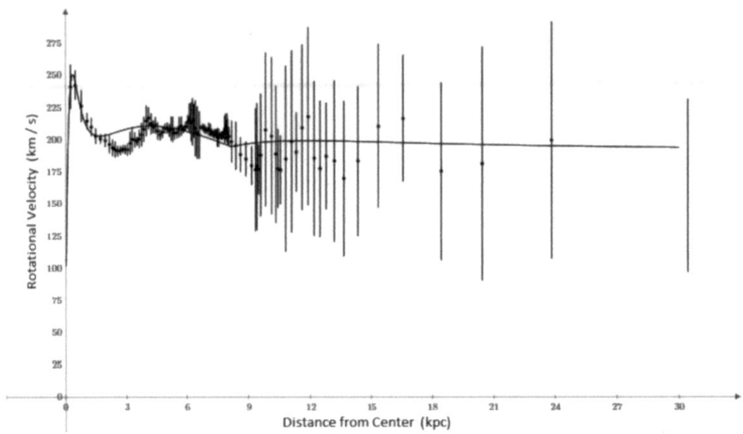

Figure 2.
The velocity vs. radial distance from the galactic center of the milky way galaxy. The data points and errors are from [46]. The solid curve is the fit to the galaxy rotation curve with the graviton model, Eqs. (13–15).

$MAE = 13.25 kms^{-1}$. For the radial distance up to r_0, the square of the correlation coefficient between the observed and modeled velocities is $r^2 = 0.136$. For the full radial distance up to 1.28 Mpc, the square of the correlation coefficient between observed and modeled velocities is $r^2 = 0.171$.

The results for NGC 3198 and the Milky Way galaxies are summarized in **Table 1**. Our graviton model of the stellar velocity data of these two galaxies exhibits a decreasing rotational velocity for increasing distance from the galactic center, which is a result of the lingering of the Newtonian velocity component in the model, using Eq. (6), expressed proportionally by

$$v(r) \propto \sqrt{\frac{1-2K_g}{r} + \frac{2K_g}{r_0}}, \tag{18}$$

for $r_0 \leq r$. This effect may mimic the external field effect observed in spiral galaxies as reported by [47]. Use of the graviton model to analyze more galaxies is necessary to quantify this effect.

For comparison, we also have modeled the stellar rotational velocity *vs* distance from the galactic centers as a version of this empirical relationship,

$$V_\theta = \frac{V_m D_L}{K_\theta + D_L} \tag{19}$$

where V_θ is the rotational velocity in the galactic plane, V_m is the average maximum velocity, D_L the observed binned distance from the galactic center, K_θ is some constant, which should not be confused with K_g. The general form of this relationship is well known in physical chemistry as the Langmuir equation

Galaxy	r_0 (kpc)	K_g	n	r^2
NGC 3198	4.254	0.421	4	0.986
Milky Way	8.157	0.418	6	0.171

Table 1.
Graviton model results from fits of Eq. (13) to the NCG 3198 and milky way full range binned data. Note that r^2 is the correlation coefficient corrected for parameter number and not the value for any radius.

Galaxy	V_m (kpc)	K_θ	r^2
NGC 3198	154 ± 2	0.05 ± 0.1	0.982
Milky Way (113)	211 ± 2	0.20 ± 0.04	0.991
Milky Way (99)	200 ± 2	-0.16 ± 0.05	0.997

Table 2.
Results from nonlinear regression of galaxy binned data with the Langmuir equation. Note that r^2 is the correlation coefficient corrected for parameter number and not the value for any radius.

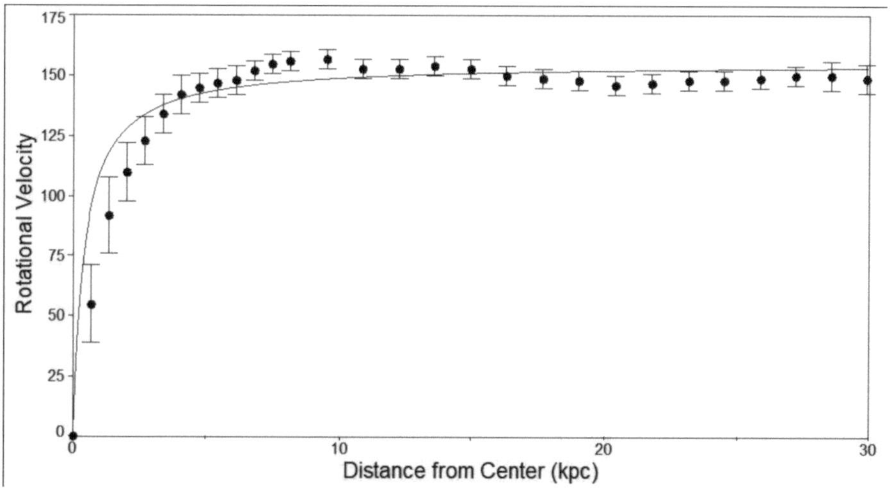

Figure 3.
The rotational velocity (km s^{-1}) vs. radial distance (kpc) from the galactic center of NGC 3198. The solid curve is a good fit to the data with the Langmuir model, Eq. (32). The errors shown here and used for modeling are the standard deviations from the observations [45].

describing a substrate-saturation curve of a binding site-limited system. Note that we allowed both V_m and K_θ to be the two free parameters. We select this model because the data exhibit an initial quick increase in V_θ followed by a long progression of stars with fairly similar V_θ as evidenced by many barred galaxies [4]. We present the results from nonlinear regression modeling Eq. (19) with the NGC 3198 and the Milky Way data in **Table 2** with all three analyses including the position of the galactic center, velocity, 0,0 without error. Note the results of Milky Way (113) including all data and the results of Milky Way (99) where the 14 data pairs closest to the galactic center are not used because the data are very noisy. This model fairly describes the Milky Way data as the values for r^2 are nearly 1.

Figure 3 presents the best fit to the binned NGC 3198 data using the Langmuir model. It appears that the Langmuir equation is a fair model to the NGC 3198 data for stars at and beyond 4 kpc from the galactic center and might be useful to estimate the average rotational velocity of stars in spiral galaxies, which are not too close to the center. Another thing to note is that the rotational velocity of the Milky Way estimated by both the graviton and Langmuir models is similar at about 200 km s^{-1}.

4. Conclusions

We suggest an alternate explanation to the usual for the Zwicky and Rubin observations without resorting to the special pleading required by the WIMP and

MACHO scenarios. Though there have been several well-intentioned attempts to observe WIMPS through many different energy ranges, we have not read of any experiment or observation that one might call success. The search continues, however, with a portion of the high energy time of the CERN facility dedicated to such experiments [48]. More astronomical observations confirming or denying, indirectly, DM are on the horizon, involving hundreds of people [49]. We wonder if the results of these will make any difference to the opinions of astronomers since observations to data of distant galaxies yielding results that both confirm and deny DM are continually reported to this day; the opinions of astronomers seem to be data-independent [50, 51].

Our results from modeling the rotational data of spiral galaxy NGC 3198 resulting from graviton physics are in close agreement to observations. Examination of **Figure 1** and the results in **Table 1** reveals our graviton model to be an excellent description of this galaxy. The results of using this model with the Milky Way data, **Figure 2**, are not nearly as good. This is probably due to two different sources. First, the Milky Way data is quite noisy, which is to be expected since astronomers are collecting information that necessarily travels through densely packed stars and gas clouds. Second, the large number of parameters needed in the mass density model to properly use our graviton model makes it very difficult to properly select the best values for modeling. A more robust mass density model is needed, rather than the rudimentary density of Eq. (16), to better match the density in the Milky Way and other complex spiral galaxies.

Our results from modeling the data from these two galaxies using the Langmuir equation are good, as seen in **Figure 3** and **Table 2**. We do not claim this model properly describes the rotational velocities of either galaxy. It may be of use to astronomers for selecting the best average velocity of stars a fair distance from the galactic center.

Although the fits to the Milky Way rotation data were weak for the graviton model, due to the poor density model, this does not diminish its potential. The graviton model is based on the physical principle that gravitons are bosons and thus should behave like photons in a gravitational field. Gravitons traveling away from a massive source should lose energy due to gravitational redshift just as observed for photons. This process takes place for gravitons within galaxies, between galaxies, and within galactic clusters and throughout the Universe. Taking proper account for gravitons can go a long way to explain dark matter and even dark energy and orbital decay of binary stars [41]. This effect use of the graviton model has been overlooked for nearly a century.

Cosmological Principle Finally, we draw the readers' attention to the commonly ignored fact that there is no evidence for WIMPS on earth (or in our solar system). The bald fact is that the properties required by DM demand WIMPS should be concentrated at the center of the earth. Few have bothered to investigate this obvious requirement; those who have made some observations have not uncovered evidence for this [52]. Dark matter proponents of WIMPS must resort to special pleading that WIMPS are "out there" but not here. This requirement is simply ignoring the cosmological principle. That is, if WIMPS are common in other galaxies including Andromeda, but not in the Milky Way, then the cosmological principle is not true. If so, astronomy is left without any principles at all.

Conflict of interest

The authors declare no conflict of interest.

Thanks

We thank K.G. Begeman for kindly allowing us use of his NGC 3198 data.

Abbreviations

BTFR	baryonic Tully-Fisher relation
CGR	Carmeli general relativity
DM	dark matter
EE	Einstein equation
MACHO	massive astrophysical compact halo object
SNe Ia	supernovae type Ia
SG	spiral galaxy
SM	standard model of particle physics
TF	Tully-Fisher (relationship)
WIMP	Weakly Interacting Massive Particles

Author details

Firmin Oliveira[1†] and Michael L. Smith[2*†]

1 Asian Observatory/James Clerk Maxwell Submillimetre Telescope, Hilo, Hawaii, United States

2 Umeå, Sweden

*Address all correspondence to: mlsmith55@gmail.com

†These idea, derivation and modeling were conceived and written by Oliveira, Smith wrote the introduction and conclusions.

References

[1] Zwicky F. On the masses of nebulae and of clusters of nebulae. The Astrophysical Journal. 1937;**86**:217-246. DOI: 10.1086/143864

[2] Rubin VC, Ford WK Jr. Rotation of the andromeda nebula from a spectroscopic survey of emission regions. The Astrophysical Journal. 1970;**159**:379-403. DOI: 10.1086/150317

[3] Rubin VC, Ford WK Jr, Thonnard N. Extended rotation curves of high-luminosity spiral galaxies. IV-Systematic dynamical properties SA through SC. The Astrophysical Journal. 1978;**225**:L107-L111. DOI: 10.1086/182804

[4] Rubin VC, Thonnard N, Ford WK Jr. Rotational properties of 21 SC galaxies with a large range of luminosities and radii, from NGC 4605 (R = 4kpc) to UGC 2885 (R = 122 kpc). The Astrophysical Journal. 1980;**238**:471-487. DOI: 10.1086/158003

[5] Persic M, Salucci P, Stel F. The universal rotation curve of spiral galaxies - I. The dark matter connection. Monthly Notices of the Royal Astronomical Society. 1996;**281**(1):27-47. DOI: 10.1093/mnras/278.1.27

[6] Hossenfelder S. Dark Matter. Or What? YouTube; 2018. Available from: https://www.youtube.com/watch?v=FN2d2cmi_Gk

[7] Cooperstock FI, Tieu S. General relativity resolves galactic rotation without exotic dark matter. arXiv. 2005. Available from: https://arxiv.org/abs/astro-ph/0507619

[8] Fuchs B, Phleps S. Comment on "General relativity resolves galactic rotation without exotic dark matter" by F.I. Cooperstock and S. Tieu. New Astronomy. 2006;**11**:608-610. DOI: 10.1016/j.newast.2006.04.002

[9] Hartnett JG. Spiral galaxy rotation curves determined from carmelian general relativity. International Journal of Theoretical Physics. 2006;**45**:2118-2136. DOI: 10.1007/s10773-006-9178-0

[10] Lelli F, McGaugh SS, Schombert JM. The small scatter of the barionic tully_fisher relation. Astrophysicsal Journal Letters. 2015;**816**(1):L14. DOI: 10.3847/2041-8205/816/1/l14

[11] Bertone G, Hooper D, Silk J. Particle dark matter: evidence, candidates and constraints. Physics Reports. 2005;**405**:279-390. DOI: 10.1016/j.physrep.2004.08.031

[12] Feng JL. Naturalness and the Status of Supersymmetry. Annual Review of Nuclear Particle Science. 2013;**63**(1):351-382. DOI: 10.1146/annurev-nucl-102010-130447

[13] Bernabei R, Belli P, Cappella F, Caracciolo V, Castellano S, et al. Final model independent result of DAMA/LIBRA-phase 1. European Physical Journal C. 2013;**73**(12):2648. DOI: 10.1140/epjc/s10052-013-2648-7

[14] Aartsen MG, Abbasi R, Abdou Y, Ackermann M, Adams J, et al. Search for dark matter annihilations in the sun with the 79-string ice cube detector. Physical Review Letters. 2013;**110**:131302. DOI: 10.1103/physrevlett.110.131302

[15] Choi K, Abe K, Haga Y, Hayato Y, Iyogi K, et al. Search for neutrinos from annihilation of captured low-mass dark matter particles in the sun by super-kamiokande. Physical Review Letters. 2015;**114**:141301. DOI: 10.1103/physrevlett.114.141301

[16] Aprile E, Aalbers J, Agostini F, Alfonsi M, Amaro F, et al. Search for

electronic recoil event rate modulation with 4 years of XENON100 data. Physical Review Letters. 2017;**118**: 101101. DOI: 10.1103/physrevlett. 118.101101

[17] Cui X, Abdukerim A, Chen W, Chen X, Chen Y, et al. Dark matter results from 54-ton-day exposure of panda X-II experiment. Physical Review Letters. 2017;**119**:181302. DOI: 10.1103/ physrevlett.119.181302

[18] Schumann M. Direct detection of WIMP dark matter: Concepts and status. Journal of Physics G: Nuclear and Particle Physics. 2019;**46**(10):103003. DOI: 10.1088/1361-6471/ab2ea5

[19] Zhou N. New WIMP and Axion Search Results From Complete Exposure of PandaX-II Released. 2020. Available from: https://pandax.sjtu.edu. cn/node/393

[20] Habermehl M, Berggren M, List J. WIMP dark matter at the international linear collider. Physical Review D. 2020; **101**:075053. DOI: 10.1103/physrevd.101. 075053

[21] Roszkowski L, Sessolo EM, Trojanowski S. WIMP dark matter candidates and searches—current status and future prospects. Reports on Progress in Physics. 2018;**81**:066201. DOI: 10.1088/1361-6633/aab913

[22] Oztas AM, Smith ML. A polytropic solution of the expanding universe-constraining relativistic and non-relativistic matter densities using astronomical results. In: Alfonso-Faus A, editor. Aspects of Today's Cosmology. IntechOpen; 2011. DOI: 10.5772/1838. ch 15

[23] Cuoco A, Kramer M, Korsmeier M. Novel dark matter constraints from antiprotons in light of AMS-02. Physical Review Letters. 2017;**118**(19):191102. DOI: 10.1103/physrevlett.118.191102

[24] Einstein A. The Principle of Relativity. Mineola, NY: Dover Pub; 1952

[25] Villanueva-Domingo P, Mena O, Palomares-Ruiz S. A brief review on primordial black holes as dark matter. arXiv. 2021. Available from: https:// arxiv.org/pdf/2103.12087.pdf

[26] Bertone G, Hooper D. History of dark matter. Reviews of Modern Physics. 2018;**90**(4):045002. DOI: 10.1103/revmodphys.90.045002

[27] Kalirai J. Scientific discovery with the James Webb Space Telescope. Contemporary Physics. 2018;**59**: 251-290. DOI: 10.1080/00107514. 2018.1467648

[28] Kasdin NJ, Bailey V, Mennesson B, Zellem R, Ygouf M, et al. The Nancy Grace Roman Space Telescope Coronagraph Instrument (CGI) Technology Demonstration. Space Telescopes and Instrumentation 2020. Optical, Infrared, and Millimeter Wave, SPIE; 2020. DOI: 10.1117/12.2562997

[29] Calcino J, Garcia-Bellido J, Davis TM. Updating the MACHO fraction of the Milky Way dark halo with improved mass models. Monthly Notices of the Royal Astronomical Society. 2018;**479**:2889-2905. DOI: 10.1093/mnras/sty1368

[30] Sorce JG, Courtois HM, Tully RB, Seibert M, Scowcroft V, et al. Calibration of the mid-infrared Tully-Fisher relation. The Astrophysical Journal. 2013;**765**(2):94. DOI: 10.1088/ 0004-637X/765/2/94

[31] Lelli F, McGaugh SS, Schombert JM. The small scatter of the barionic Tully–Fisher relation. Astrophysical Journal Letters. 2016;**816**(1):L14-L19. DOI: 10.3847/2041-8205/816/1/l14

[32] Feynman R, Morinigo F, Wagner W, Hatfield B, Pines D.

Feynman Lectures on Gravitation. Taylor & Francis eBooks: CRC Press; 2002

[33] Jungman G, Kamionkowski M, Griest K. Supersymmetric dark matter. Physics Reports. 1996;**267**:195-373. DOI: 10.1016/0370-1573(95)00058-5

[34] Kim JE, Alverson G, Nath P, Nelson B. A review on axions and the strong CP problem. AIP Conference Proceedings. 2010;**1200**:83. DOI: 10.1063/1.3327743

[35] Carvalho A. Gravity particles from warped extra dimensions, predictions for LHC. arXiv. 2014. Available from: https://arxiv.org/pdf/1404.0102.pdf

[36] Kroupa P. Galaxies as simple dynamical systems: observational data disfavor dark matter and stochastic star formation. Canadian Journal of Physics. 2014;**93**(2):169-202. DOI: 10.1139/cjp-2014-0179

[37] Ghag C. The dark matters with Chamkaur Ghag. Online lecture June 2021. Available from: https://www.newscientist.com/science-events/dark-matters/

[38] Hossenfelder S. Are Dark Matter and Dark Energy Scientific? Youtube; 2020 Available from: https://www.youtube.com/watch?v=GatiekRziok

[39] Milgrom M. MOND vs. dark matter in light of historical parallels. Studies in History and Philosophy of Science Part B: Studies in History and Philosophy of Modern Physics. 2020;**71**:170-195. DOI: 10.1016/j.shpsb.2020.02.004

[40] Hossenfelder S. The Situation has Changed. Youtube; 2021 Available from: https://www.youtube.com/watch?v=4_qJptwikRc

[41] Oliveira FJ. The principle of equivalence: periastron precession, light deflection, binary star decay, graviton temperature, dark matter, dark energy and galaxy rotation curves. JHEPGC. 2021;7(2). DOI: 10.4236/jhepgc.2021.72038

[42] Milgrom M. A modification of the Newtonian dynamics as a possible alternative to the hidden mass hypothesis. The Astrophysical Journal. 1983;**270**:365-370. DOI: 10.1086/161130

[43] Carmeli M. Relativity: Modern Large-Scale Spacetime Structure of the Cosmos. Singapore: World Scientific Pub. Co. Pte. Ltd; 2008. pp. 242-247

[44] McGaugh SS. The baryonic Tully-Fisher relation of galaxies with extended rotation curves and the stellar mass of rotating galaxies. The Astrophysical Journal. 2005;**632**(2):859-871. DOI: 10.1086/432968

[45] Begeman KG. HI Rotation Curves of Spiral Galaxies [thesis]. Groningen, Netherlands: Astronomy, University of Groningen; 2006. Available from: https://pure.rug.nl/ws/portalfiles/portal/2841676/c4.pdf

[46] Sofue Y. Grand rotation curve and dark-matter halo in the milky way galaxy. PASJ. 2012;**64**(4):75. DOI: 10.1093/pasj/64.4.75

[47] Chae KH, Lelli F, Desmonnd H, McGaugh SS, Li P, Schombert JM. Testing the strong equivalence principle: Detection of the external field effect in rotationally supported galaxies. The Astrophysical Journal. 2020;**904**(1). DOI: 10.3847/1538-4357/abbb96

[48] CERN Cornering WIMPs with ATLAS. 2020. Available from: https://cerncourier.com/a/cornering-wimps-with-atlas/

[49] Abdallah H et al. Search for dark matter annihilation in the Wolf-Lundmark-Melotte dwarf irregular galaxy with H.E.S.S. Physial Review D.

2021;**103**(2021):102002. DOI: 10.1103/PhysRevD.103.102002

[50] de Martino I et al. Dark matters on the scale of galaxies. Universe. 2020;**6**: 107. DOI: 10.3390/universe6080107

[51] Shen Z et al. A tip of the red giant branch distance of 22.1 1.2 Mpc to the dark matter deficient galaxy NGC 1052–DF2 from 40 orbits of hubble space telescope imaging. Astrophysical Journal Letters. 2021;**914**:L12. DOI: 10.3847/2041-8213/ac0335

[52] Kunnen J. A search for dark matter in the center of the earth with the ice cube neutrino detector [thesis]. Ixelles, Belgium: Vrije Universiteit Brussel; 2015. Available from: https://iihe.ac.be/sites/default/files/thesis-jan-kunnen-icecube-phd-2015pdf/thesis-jan-kunnen-icecube-phd-2015.pdf

Chapter 4

The Most Probable Cosmic Scale Factor Consistent with the Cosmological Principle, General Relativity and the SMPP

Arthur N. James

Abstract

Current literature on the evolution of the cosmic scale factor is dominated by models using a dark sector, these all involve making many conjectures beyond the basic assumption that the Cosmological Principle selects a space–time metric of the Friedmann–Lemaître–Robertson–Walker type through which ordinary Standard Model of Particle Physics matter moves according to General Relativity. In this chapter a different model is made using the same basic assumptions but without making extra conjectures, it depends on following the idea introduced by Boltzmann that when physically meaningful concepts fluctuate the value which will be observed is the one which has the highest probability. This change removes the mathematically incorrect procedure of averaging the matter density before solving Einstein's Equation, the procedure which causes the introduction of many of the conjectures. In the non-uniform era the changes are that the evolution of the scale factor is influenced by the formation of structure and removes the conjecture of having to use two inconsistent probability distributions for matter through space, one to calculate the scale factor and one to represent structure. The new model is consistent from the earliest times through to the present epoch. This new model is open and matches SNe 1a redshift data, an observation which makes it a viable candidate and implies that it should be fully investigated.

Keywords: cosmology, gravitation, dark matter, dark energy

1. Introduction

This first section introduces the motivational background to the study described in this chapter. The study is a response to the difficulties found by an academic physicist trying to upgrade from an amateur cosmologist, who just followed conclusions published in the scientific literature, to a more professional stance by finally studying General Relativity late in life. In 1956 before starting nuclear physics research for a doctoral degree a lifelong interest in the cosmos was triggered by Martin Ryle's course in Radio Astronomy where he described how using Malmquist bias on the C2 Catalogue source counts he could demonstrate that the cosmos was evolving [1].

The teaching of physics in the early 1950s was largely in the style of natural philosophy which means that nature was observed and then modelling of the phenomena was made by searching for some appropriate mathematics. The current position in cosmology is different. For many years through the middle of the twentieth century the simple assumption of a flat space Friedmann–Lemaître–Robertson–Walker (FLRW) metric was used despite the absence of any direct observational evidence supporting this choice. Observations of individual objects in the sky are satisfactorily described using such flat space but there are unobservable consequences such as horizons present in the associated cosmology. Later the mathematical invention of Cosmic Inflation to overcome the horizon problem associated with the flat cosmologies appears to have converted the flatness assumption into an approved folk lore not to be questioned.

Since that time cosmology has included many conjectures required to match the real observed Universe, each of these should carry with it an unknown improbability weighting. Because of the accumulation of such weightings the old fashioned way of choosing between different models describing the same observations would have been to quote Occam's Razor and select the model with the least conjectures so as to improve the odds of being correct.

The study described here is an attempt to use only well authenticated physics and observations in a return to basics and a natural philosophers method for constructing a model for the evolution of the cosmos. This leads to a new model for the cosmological scale factor which is essentially free from additional conjectures.

2. Background

A hundred years ago Friedmann combined the cosmological principle with Einstein's Equation to predict ways in which a cosmic metric could change with time, his initial model filled the cosmos with a uniform non-relativistic distribution of matter which slowed down any existing expansion of the cosmos. The symmetry described by the cosmological principle enables the modelling of an expanding evolving cosmos with curved space–time because these conditions imply that the background space–time metric must be an FLRW metric. At low matter densities Friedmann's solution has open space sections, as the density increases the solution appears to change smoothly through one special solution with a flat space section into the high density region where the space sections are closed and the Universe collapses back to a point. That description is misleading, in more general situations when other fluids are also present in the cosmos there are three disconnected families of solutions, open, flat and closed. Within each family there are many variations in the way that the cosmic scale factor can change with time depending on the particular mixture of substances filling the cosmos. Against any assumed background metric for the Universe the Universe's content of ordinary matter can be modelled through the formation of structures using conventional physics, see e.g. Peebles' textbook [2].

The present epoch of the Universe is characterised by a non-uniform distribution of matter and a simple Friedmann solution is not possible because Einstein's Equation is non-linear and the procedures of solving and smoothing must be carried out in that order, they do not commute (see e.g. page 452 in Padmanabhan's book [3]). When the wrong order of procedures is used the source distribution does not properly represent the natural distribution so that the solution to Einstein's Equation is that for a non-existing situation, it must be nonsense in the context of representing nature. The Dark Sector cosmologies which are currently widely used incorporate this mathematical misdemeanour, the cosmology becomes the solution to a problem where its dust component has to be in two places at once, both in galaxy clusters and simultaneously everywhere else, this property of the dust material defies relativity. Being aware of these difficulties suggests questioning all the conjectures which form the essential starting point for Dark Sector models.

The study described here has two themes, one is the blunt rejection of the Dark Energy models, the second theme is a proposal using only well established physical concepts that an open cosmology is highly probable and should be examined further by groups with the appropriate knowledge, skills and computing resources.

In Section 3 the essential physical knowledge and observations which provide a common basis of knowledge for both the new and the Dark Sector models is described, this knowledge is used to make clear those situations where an additional conjecture is necessary to make further progress. All of the science used is well described in many textbooks, the ones quoted here are Padmanahaban's [3] and Peebles' [2]. The notation commonly used in applying such basic knowledge to the formation of cosmological models is introduced by describing the uniform radiation era. Section 4 will introduce the proposed new method using probability density distributions using a simple description of the present epoch, the resulting model is an open cosmology. Section 5 describes how the new method may also be used to describe the whole of the evolution of the cosmic background metric from the radiation era to the present day. Section 6 indicates many of the conjectures which have to be made to construct the Dark Sector models. Section 7 concludes by advocating development of the open model and lists some questions which have to be addressed and answered before continuing to use the Dark Sector models.

3. Basic knowledge common to all models

The symmetry described by the cosmological principle implies that the cosmos has a background space–time metric of the FLRW type. The three families of these metrics are distinguished by their space sections being either open, closed or flat, the cosmic scale factor of these metrics will be increasing with time to describe the expansion. The 3D curvature of the space of each time slice is determined by the geometry of the FLRW metric, the flat space section family have infinite radius of curvature whereas the open and closed families have a radius of curvature equal to the scale factor, see e.g. Padmanabhan [3]. This mathematical fact is currently being widely ignored in fitting procedures such as that of the Planck collaboration [4] where small deviations from "flatness" are being interpreted as indicating non flat space sections. Such small curvatures would indicate departures from the cosmological principle which is the most essential assumption of the entire conception of modelling background space-times.

Observations and laboratory experiments confirm that the only directly observable substances in the Universe are made from components of the Standard Model of Particle Physics (SMPP). If in addition to the cosmological principle the contents of the universe are also uniform across each time slice then Einstein's Equation of General Relativity can be used to determine the way in which the cosmic scale factor changes with time. There are two types of fluid which can be made from the particles of the SMPP, particles which are moving with relativistic speed make a radiation fluid with significant pressure whereas particles moving slowly make a dust fluid with zero pressure. The equation of state relation between the energy density and pressure influences the rate of change of the cosmic scale factor through Friedmann's Equations which combine General Relativity and FLRW metrics.

The energy density and pressure of particulate fluids are thermodynamic intensive quantities and will be fluctuating, they should correctly be described using the statistical methods of physics and probability distributions. The convention for pressure and temperature introduced by Boltzmann but now used throughout physics for fluctuating quantities is to use the most probable value as the value to represent overall behaviour with the fluctuations being assumed relative to that value. If electromagnetic signalling is used as an example then two situations must be considered, when many particles are involved the fluctuations are thermal noise, symmetrical around the mode of the distribution which equals the average, but when only very small numbers are involved the fluctuations are shot noise with the mode being zero and with fluctuations in just one direction. Probability distributions where the most probable value is zero will be used in this study when considering non-uniformity such as the matter density in the Universe in the present epoch.

Through the second half of the twentieth century it became apparent that the early universe could be modelled using a radiation fluid uniformly filling an expanding cosmos, the radiation era of the Big-Bang. The cosmic density of light elements predicted by Big-Bang nucleosynthesis verifies both the expansion rate as being that of a radiation era and that the density of SMPP matter is close to the value estimated from observations of astronomical objects. The importance of the Cosmic Microwave Background Radiation (CMBR) to cosmology cannot be exaggerated, it is the most reliable demonstration of the cosmological principle whilst simultaneously justifying the assumption that before its emission at the recombination temperature for atomic Hydrogen the density distribution was uniform. The fluctuations in the CMBR imply that the cosmological principle must be used in a form where the density probability distribution is the same at every spacial point in a particular time slice.

The CMBR is an important boundary in the development of cosmic expansion, before that time uniformity was ensured by radiation mixing throughout the ionised plasma. The emission of the CMBR signals the end of ionisation and the cessation of the mixing, the universe becomes non-uniform. That the universe is non-uniform is obvious in the present epoch, the cosmos is populated by many galaxies in complex structures, the way in which these can form from the minute fluctuations in the CMBR can be explained using ordinary physics [2].

Towards the end of the uniform era the dominating fluid of SMPP particles filling the universe changes in character from the earliest relativistic radiation fluid to a non-relativistic dust fluid with zero pressure. Using the normal Friedmann description the equation showing the relationship between Hubble's constant $H(t)$ and the cosmic scale factor $a(t)$ adjusted so that these take the values H_o and $a_o(t) = 1$ at the present time is

$$H^2 = H_o^{\,2}\left[\Omega_R a^{-4} + \Omega_{BM} a^{-3} - kc^2 \left(H_o a\right)^{-2}\right] \tag{1}$$

The Ω_R and Ω_{BM} are cosmic densities of radiation and baryonic matter normalised in the usual way to that of the flat Friedmann matter only cosmological model. The curvature term where k equals -1, 0 or + 1 has a different character, it is part of the FLRW geometry and represents the relation between the absolute value of the cosmic scale factor and the spacial curvature of the particular chosen geometry, it has nothing to do with Einstein's Equation but stems only from the symmetry of the Cosmological Principle [3].

At early times in the expanding universe when the cosmic scale factor is smallest the radiation density Ω_R term dominates but when the baryonic matter density Ω_{BM} deduced from the Big-Bang nucleosynthesis is used the non-relativistic matter term will have just become relevant at the emission of the CMBR and a simple Friedmann matter only model suggests an open cosmos. The curvature term will always be insignificant during the uniform era which is therefore insensitive to its value and whether k is -1, 0 or + 1 can not be deduced from observations of the uniform era.

4. The present epoch and the open model

Observations of the local environment have shown that it is characterised by an array of galaxies arranged in structures which are separating according to the Hubble flow. At each point in space Einstein's Equation ensures that it is only the local density of matter which will control the rate of change of volume of a small element of space around that point, the local matter density changes the local scale factor. Remote massive objects will distort the space element as they approach but will return it to its original state after they have passed, not affecting the local scale factor. The matter density probability distribution at each point can be estimated by considering the density distribution over space, it is peaked strongly at zero. The vast majority of points have zero density which means that the local scale factor will be expanding progressively faster than anywhere where the matter density is positive. It is this mechanism where dense regions expand more slowly than empty regions which is essential to trigger the formation of structure. However the main consequence in the context of the local cosmology is that emptiness is becoming ever more frequent. Using a maximum probability algorithm that the best estimate of the cosmic scale factor must be the most frequent local scale factor sets the cosmic scale factor as that for emptiness.

The FLRW metric for empty space is Friedmann's well known empty universe solution, a useful demonstration of this is given by Vishwakarma [5], he also shows that this metric and a Dark Sector model metric have equally good fits to SNe 1a redshift data. The empty universe solution has a metric like that of the cosmology proposed by Milne [6] in 1935, open and expanding with its cosmic scale factor being the product of the velocity of light and the age of the universe. The galaxies are in free fall and simply drift apart making the Hubble flow. At any time Hubble's constant is the inverse of the age of the universe and its present value is the only parameter of this cosmology. The obvious conclusion must be that Milne's metric, a good solution of Einstein's Equation of General Relativity and the Cosmological Principle, is the best metric with which to approximately model the present day epoch.

In this argument for the present epoch it is assumed that Einstein's Equations have been exactly solved by Nature before smoothing the solution is attempted, it is then completely free from the commutation misdemeanour inherent in every dark energy model. The resulting model, Milne's metric, is just a normal FLRW model of a smooth universe neglecting blemishes such as the contamination by many massive galaxies, a situation which also occurs for all other models. Einstein's Equation is fully respected in this open model, both in the scale factor solution and in the next weak curvature approximation which has to be made to describe the interrelated motions of all the massive objects through the use of Newton's Laws of Motion.

5. The open model from the radiation era to the present day

If the mode of the density probability distribution rather than the average is used for the uniform era then modelling the scale factor using Eq. 1 is unaffected because the uniformity ensures only a narrow distribution of densities and the mode of the density distribution will be almost identical to the mean in this case. After the emission of the CMBR the density distribution becomes non-uniform and changing with time. Using the mode of the density distribution for the non-uniform era produces a model where the cosmic scale factor responds to the formation of structure by modifying the Friedmann equation as time evolves.

The model is generated by imagining a stepwise numerical time integration technique where the cosmic scale factor change through a step will be calculated using Friedmann's Equation with the density mode from the start of the step while the changes to the density distribution function during the step are calculated non relativistically in the usual way. At the end of the step the density mode will be different so the Friedmann's Equation which should be used in the next step to compute the scale factor will be different. In this way the initial conditions for each step match the solution and the nearness of the approximation to a continuous integration with smoothing following solution will increase as the step size decreases. An idiotically simple example of using the wrong order of procedures in a non-linear problem and then applying such an integration procedure to evade its effects is given as an appendix.

In this stepwise process the Friedmann equation to be used is shown below including explicitly the curvature term for an open cosmology (where $Ha = c$)

$$H^2 = H_o^{\,2}\left[\Omega_R a^{-4} + \Omega_{BMC}(t)a^{-3} + c^2\left(H_o a\right)^{-2}\right] \qquad (2)$$

The term $\Omega_{BMC}(t)$ represents that Baryonic Matter Component which affects the cosmic scale factor at each time, this term changes from its full value Ω_{BM} at the CMBR time to zero for the present day the details depending on how the mode of the matter distribution function evolves with time. Because Ω_R and $\Omega_{BMC}(t)$ are small for the present epoch the curvature term $c^2(H_o a)^{-2}$ for the Milne metric dominates. In any fitting procedure for the CMBR and structure formation the only parameters will be the properties of the CMBR fluctuation distribution, the Hubble constant H_o and the baryonic matter content Ω_{BM}, all parameters having clear physical meaning in relation to the SMPP and General Relativity and related to observations.

If the early value for Hubble's constant predicted from the CMBR does not match the late value determined by the SNe 1a data then the value for the matter term required to fit the data may be larger than Ω_{BM}, perhaps dark matter such as that suspected from studies of the Bullet cluster [7] will have to be included. This dark matter must behave in the same way as SMPP matter according to the rules of General Relativity and should not be considered in any way as being similar to the cosmic dust component of Dark Energy models.

A qualitative description of what happens as the Cosmic Baryonic Matter term reduces to zero is simply that isolated discrete lumps of matter are unable to influence the scale factor of the whole of a time slice at once, it should be obvious that such an influence violates relativity principles. As stars and galaxies form they remove the matter from its cosmic role leaving the smaller curvature term to finally take over control of the expansion.

6. The problems of the concordance model

The simplest Dark Energy models introduce four conjectures into their model for the cosmic scale factor, flat space sections, cosmological Dark Matter (DM),

Dark Energy (Λ) and weird properties for the dust component. The Friedmann equation on which all these models are based is shown here with a zero a^{-2} curvature term because of the flatness assumption

$$H^2 = H_o^{\,2}\left[\Omega_R a^{-4} + (\Omega_{BM} + \Omega_{DM})_{WD}\, a^{-3} + \Omega_\Lambda\right] \quad (3)$$

In this equation both matter terms, baryonic Ω_{BM} and dark Ω_{DM}, have been bracketed together because the physical properties for this Weird Dust (WD) are very strange. The Weird Dust appears to have two different density distributions across a time slice, the modeller chooses to use one or the other depending on context, a smooth distribution to compute the cosmic scale factor evolution but then a non-uniform one to compute structure formation. This inconsistency within the model makes it bad science due to the illogicality where the premise of uniformity for the matter content does not match the outcome of the model calculations which predict the destruction of uniformity.

Another way of describing the situation is that in addition to the normal physical properties of an isolated concentrated lump of dust these models conjecture an extra weird property for that isolated concentrated lump of dust, that of being able to instantly affect the universe's cosmic scale factor uniformly throughout the universe. Such behaviour is not allowed by General Relativity which respects the restrictions of a finite velocity of light.

Each of the conjectured Dark components comes with a quantity of substance and an equation of state all of which are artefacts of the model. These two quantities and two functions provide enough flexibility in the fitting procedures for the solution to respond to the attraction towards flatness imposed on the models by using a zero cosmic curvature term in the Friedmann equation. The quantity 'Ω_k' = $(1 - \Omega_M - \Omega_\Lambda)$ which is seen for example in the fitting procedures for Planck data [4] cannot represent curvature in a FLRW metric, its value is generally found to be near zero which must simply be a measure of the accuracy within which the fit has approached flatness.

One more conjecture is made in Dark Energy models because the horizon introduced by the flatness means that Inflation must also be conjectured to ensure the cosmological principle is present through both the uniform and non-uniform eras.

7. Summary and conclusion

A new technique to find an approximate solution to Einstein's Equation for the cosmic scale factor in a non uniform universe has been found by going back to basic physics, SMPP plus General Relativity, and following Boltzmann's use of maximum probability concepts in physics. This technique causes the cosmic scale factor to be affected by the formation of structures in the universe. The resulting open cosmological model is very different from conventional cosmologies, its simplest predictions are that evolution from the early radiation era to the present epoch produces an empty cosmos with a small contamination of massive galaxies drifting apart in accordance with the observed SN1a redshift data. There appears to be no obvious contradictory observational data to this new cosmological model. This open cosmology has a lower density than flat models, the expansion through the CMBR epoch and the structure forming era will be slower giving extra time for the formation of early astrophysical objects. Detailed structure calculations to see how this open model fits the CMBR and the Baryonic Acoustic Oscillation data are required to validate it further, computations which can be justified by the simplicity of all the physical concepts required to establish it.

The empty universe's expanding cosmic scale factor can only be modelled by an open cosmology implying that the cosmos has always been open and causally connected. A causally connected model of the universe does not require inflation to establish the uniformity of the cosmological principle. Such a model implies little about the most primordial universe, it must be open, contain the SMPP and a radiation spectrum to match the details of the CMBR.

Examination of Dark Energy models of the cosmos suggests several questions about the conjectures used in the models which should be answered successfully before proceeding to further investigations. The main objection to the Dark Energy models must be the mathematical commutation of procedures misdemeanour, that this is a problem has not gone unacknowledged but the problem has not been confronted directly, it has only been circumvented by additional conjectures which do not eliminate the misdemeanour (see [8] and references therein). Apart from this misdemeanour but perhaps in consequence of the misdemeanour a list of questions requiring answers are:-.

- Why select a flat FLRW metric, what model independent observation supports such a choice?
- There may be dark matter such as that which might be concentrated in objects such as the Bullet Cluster but what is the conjectured uniformly distributed cosmic dark matter?
- What is the meaning of the strangely weird conjectured properties of the dust component in the concordance model Friedmann equation?
- Where does the eternal supply of energy for the conjectured dark energy come from?

Appendix on solving and smoothing in non-linear systems

The Dark Energy models for the cosmic scale factor average the source term in Einstein's Field Equations before solving, but that problem is non-linear meaning that a mathematical misdemeanour has been included right at the beginning of the modelling, the wrong initial conditions have been used to solve the problem and the answer must be wrong. A peculiar feature of this problem is that the equations are correct and nature provides the correct solution, a consequence is that the Dark Energy models introduce artefacts to correct the wrong solution towards the correct solution, these are Dark Matter and Dark Energy. The following idiotically simple problem is given here to illustrate how this has been done.

Consider a problem where the answer is known from other considerations to be 1/3, the question is:-.

"What is the average value of y where $y = x^2$ over the range $-1 < x < +1$?"

Averaging then solving gives the answer as zero, wrong because solving and averaging are performed in the wrong order. The initial conditions have been altered from the correct situation to an incorrect one. It is possible to adjust this answer by conjecturing an arbitrary parameter which can be adjusted to match the known answer. This arbitrary parameter will be adjusted to the value 1/3, this parameter is an artefact of the method used to solve the problem but has no real meaning.

Now split the range up into small segments, averaging first for each segment then solving, a final averaging of all the intermediate steps gives an answer quite close to the previously known correct value. By responding to the changing situation as the value of x increases the error has been vastly reduced and no arbitrary constant is required. The error will depend linearly on the step size enabling its size and effect to be detected.

In modelling the development of structure in cosmology the solution is too complicated for normal integration but essentially Dark Energy models do the whole problem without acknowledging the changing situation. The models have to introduce artefacts such as extra dust and Dark Energy, with arbitrary constants and functions, in order to fit observations. The new method using the mode of the matter probability distribution leading to the open cosmology will have reduced the effect of the mathematical misdemeanour by responding to the changing situation, the smaller the step size the smaller the error. No extra artefacts will have to be introduced.

Author details

Arthur N. James
Department of Physics, University of Liverpool, UK

*Address all correspondence to: anjames@ns.ph.liv.ac.uk

References

[1] Ryle, M. The Observatory. 1955; **75**: 137

[2] Peebles P. J. E., Principles of Physical Cosmology. Chichester: Princeton University Press; 1993.

[3] Padmanabhan T., Gravitation, Foundations and Frontiers. Cambridge: Cambridge University Press; 2010.

[4] Ade P. A. R. et al, Planck 2015 results, XIII. Cosmological parameters. Astron. Astrophys, 2016; **594 A13**

[5] Vishwakarma R. G., Mysteries of the Geometrization of Gravitation. Res. Astron. Astrophys. 2013; **13:** 1409-1422

[6] Milne E. A., Relativity, gravitation and world-structure. Oxford: Clarendon Press; 1935

[7] *Clowe D., Gonzalez A. and Markevich M., Weak-Lensing Mass Reconstruction of the Interacting Cluster 1E 0657-558: Direct Evidence for the Existence of Dark Matter. Astrophys. J. 2004;* **604** *(2): 596-603*

[8] Buchert T., Mourier P. and Roy X., On average properties of inhomogeneous fluids in general relativity III: general fluid cosmologies. Gen. Rel. Grav; 2020: **52** 27

Chapter 5

Black Holes as Possible Dark Matter

Aloke Kumar Sinha

Abstract

Black holes and Dark matter are two fascinating things that are known very little. They may have non gravitational interactions, but those are definitely extremely feeble in comparison to their gravitational interactions. Nowadays some people think that one may contain the other. In this chapter we will see that some black holes may contain the dark matter. These black holes decay under Hawking radiation, but do not vanish completely. They produce stable end states due to both quantum gravitational effects and thermodynamic reasons. These end states are the replicas of what we call dark matter. We will develop the complete theory for decay of such black holes, starting from some scheme independent assumptions for the quantum mechanical nature of the black holes. We will then consider explicit examples of some black holes to show that they indeed produce replicas of dark matter at their end states. Thus this chapter is going to be a manuscript for theoretical development of black hole decay from a quantum mechanical perspective and its consequences for producing replicas of dark matter.

Keywords: Quasi thermal stability, Thermal black holes, Black hole phase transition, Quantum gravity, Dark matter

1. Introduction

Einstein had first shown, with the help of his classical field equations of general theory of relativity, that black holes accreted everything surrounding them [1, 2]. Hence they are expected to grow in size in an unbounded manner. His theory was entirely classical. But Hawking later invoked quantum mechanics in the context of black hole [3], to study its interaction with matters surrounding it. He proved explicitly that black holes could radiate and as a consequence they decayed away. Thus a black hole radiates along with simultaneous accretion.

Hawking considered only matters as quantum entities, but spacetime was still classical in his theory. Hence in his theory, black holes were still classical. Thus this theory was semi classical as matters were treated differently in comparison to black holes. We had resolved this issue in our earlier works [4, 5]. Semiclassical analysis claimed the thermal instability of asymptotically flat, non extremal black holes under Hawking radiation. They are unstable as their specific heat is negative [6, 7] and have been deduced from semiclassical facts based on their classical metric. Their temperature increases as they lose mass, indicating a complete thermal run away process. It is to be noted that semiclassical analysis explicitly depends on the classical metric of a black hole. Hence it is inherently a 'case-by-case' analysis. This shortcoming implies that such semiclassical analysis cannot give general results about the thermal stability of generic black holes under Hawking radiation. Semiclassical analysis predicted the thermal instability of asymptotically flat black holes from the negativity of their specific heat, defined semi classically from their metric. But this result does not say anything in general about an arbitrary black hole. It is of course true that gravity is yet to be quantized fully. But we realistically expect certain symmetries for that theory [4]. These symmetries are sufficient for us to construct the grand canonical partition function of a generic black hole, if we assume the black hole to be in contact with the rest of the universe, that acts as a heat bath. We derived the criteria for thermal stability of a generic black hole with arbitrary number of parameters in any dimensional spacetime, based on the convergence of the grand canonical partition function [5]. These criteria appeared as a series of inequalities, connecting second order derivatives of black hole mass with respect to its parameters.

These criteria imply that AdS black holes with fixed cosmological constant are stable under Hawking radiation for a certain range of their parameters [4]. We have also noticed that asymptotically flat rotating charged black holes satisfy some of the stability criteria, but not all together, in certain regimes of spacetime [4, 5]. Thus although these black holes decay, they are different from unstable black holes, like asymptotically flat Schwarzschild black holes. These black holes are named as "Quasi Stable" black holes. We will later see that AdS black holes with varying cosmological constant are also quasi stable under Hawking radiation.

We had calculated the fluctuations for the parameters of a stable black hole and they were expectedly turned out to be very small [8]. These tiny fluctuations are actually the indications of the stability for a black hole. We did the same for quasi stable black holes and it resulted in tiny fluctuations for some parameters [9], like stable black holes, in a certain regime of parameter space. This is as quasi stable black holes satisfy some of the stability criteria. This makes them slow down their decay rate in certain regimes of their parameter space [9].

Black holes, like ordinary thermodynamic systems, also have different phases. Stable and unstable black holes respectively possess stable and unstable phases in possible allowed regimes of their parameter spaces. The respective examples are AdS black holes with fixed cosmological constant and asymptotically flat Schwarzschild black holes. Unstable black holes remain in the same phase during their decay. Stable black holes likewise stay in a stable phase, maintaining equilibrium with their surroundings and hence they do not decay under Hawking radiation. But things are changed entirely for quasi stable black holes. We had already shown that quasi stable black holes also have various different phases. The quasi stable black holes undergo phase transitions among these phases during their decay process. The nature of fluctuations change from one phase to another phase. In this way quasi stable black holes decay under Hawking radiation. But at the end states, most of these black holes become tiny balls of the order of Planck size. They settle down to these tiny size balls due to quantum gravity effects. On the other hand, some other parameters of certain quasi stable black holes settle down to their macroscopic values at the end states. Thus we see that these black holes become thermodynamically stable, preventing further decay under Hawking radiation. Hence they stop interacting with the rest of the universe, except gravitational interaction. Thus these black holes seem to behave like dark matter (the way we call it). In fact some of these black holes may have electric charge as well. Hence it may correspond to charged dark matter. But some unknown mechanism must be there to prevent it from interacting with the universe through known electrical interaction.

This chapter is organized as follows: A detailed discussion on thermal stability of black holes is done in Section 2. In the next section, we have discussed quasi stability and phase transitions of quasi stable black holes. In the following section, we have considered some examples of quasi stable black holes and have discussed their quasi stability and hence the possible connection with dark matter. We finished in the next chapter with a special note.

2. Thermodynamic stability criteria for black holes

A rotating, electrically charged black hole is represented classically by four parameters (M, Q, J, A), where M, Q, J, A are respectively the mass, electric charge, angular momentum and horizon area of the black hole. These four quantities are related by a relation on the horizon. Thus these parameters are expected to be promoted as operators if black hole can be treated as a quantum system. Three out of these four parameters are independent and the remaining one depends on the other three. It is certainly not possible to have charged rotating black hole without any mass and horizon area. Thus $\hat{Q}$ and $\hat{J}$ have to play the status of primary operators i.e. role of fundamental observables. We choose $\left(\hat{A}, \hat{Q}, \hat{J}\right)$ to be the primary operators and $\widehat{M}$ to be the secondary operator. Hence $\widehat{M}$ as an operator becomes $\widehat{M} = \widehat{M}\left(\hat{A}, \hat{Q}, \hat{J}\right)$. Now horizon area, like electric charge, is invariant under SO(3) rotations beside its invariance under U(1) gauge transformation. $SO(3)$ generates angular momentum while global gauge group $U(1)$ generates electric charge. These give the following commutation relations,

$$\left[\hat{A}, \hat{J}\right] = \left[\hat{A}, \hat{Q}\right] = \left[\hat{Q}, \hat{J}\right] = 0 \tag{1}$$

Since $\widehat{M}$ is a quantum operator of secondary observable $(M(A, J, Q))$, Eq. (1) can be extended as,

$$\left[\hat{A}, \hat{J}\right] = \left[\hat{A}, \hat{Q}\right] = \left[\hat{A}, \widehat{M}\right] = \left[\hat{Q}, \hat{J}\right] = \left[\widehat{M}, \hat{Q}\right] = \left[\hat{J}, \widehat{M}\right] = 0 \tag{2}$$

Thus $\hat{Q}, \hat{J}, \hat{A}$ can have simultaneous eigenstates. Hence definite values of electric charge, angular momentum and horizon area can be assigned to a black hole up to quantum and thermodynamic fluctuations. The eigenvalues of $\hat{Q}, \hat{J}$ and $\hat{A}$ are precisely the parameters used in the classical metric of a black hole to express its mass (M) as a function of them. We consider the isolated horizon to be the boundary of the black hole.

2.1 Quantum geometry

The boundary degrees of freedom and their dynamics of a classical spacetime is determined by the boundary conditions. For a quantum spacetime, fluctuations of the boundary degrees of freedom have a 'life' of their own [10, 11]. Hence the Hilbert space of a quantum spacetime with boundary has the tensor product structure $\mathcal{H} = \mathcal{H}_b \otimes \mathcal{H}_v$, where b (v) denotes the boundary (bulk) component.

So a generic quantum state$(|\Psi\rangle)$ is expandable as,

$$|\Psi\rangle = \sum_{b,v} C_{b,v} |\chi_b\rangle \otimes |\psi_v\rangle \tag{3}$$

where, $|\chi_b\rangle$ and $|\psi_v\rangle$ are respectively the boundary and bulk component of the full quantum state.

The total Hamiltonian operator($\hat{H}$) is given as,

$$\hat{H} \equiv \left(\widehat{H_b} \otimes I_v + I_b \otimes \widehat{H_v}\right) \tag{4}$$

where, respectively, $\widehat{H_b}\left(\widehat{H_v}\right)$ are the Hamiltonian operators on $\mathcal{H}_b(\mathcal{H}_v)$ and $I_b(I_v)$ are the identity operators on $\mathcal{H}_b(\mathcal{H}_v)$.

In presence of rotation and electric charge, $|\psi_v\rangle$ is be the composite bulk state and consequently it is annihilated by the full bulk Hamiltonian i.e.

$$\widehat{H_v}|\psi_v\rangle = 0 \tag{5}$$

This is the quantum analouge of the classical Hamiltonian constraint [12].

The charge operator ($\hat{Q}$) is defined as,

$$\hat{Q} \equiv \left(\hat{Q}_b \otimes \hat{I}_v + \hat{I}_b \otimes \hat{Q}_v\right) \tag{6}$$

where, $\hat{Q}_b$ and $\hat{Q}_v$ are respectively the charge operators for the boundary($|\chi_b\rangle$) and the bulk states ($|\psi_v\rangle$).

Electric charge is defined on the horizon of a classical black hole (e.g. Einstein-Maxwell or Einstein-Yang-Mills theories in [13]) and hence bulk does not carry anything i.e. $Q_v \approx 0$, the Gauss law constraint for electrodynamics. Hence, its quantum version takes the form,

$$\hat{Q}_v|\psi_v\rangle = 0 \tag{7}$$

Similarly angular momentum operator ($\hat{J}$) is defined as,

$$\hat{J} \equiv (\hat{J}_b \otimes \hat{I}_v + \hat{I}_b \otimes \hat{J}_v) \tag{8}$$

where $\hat{J}_b$ and $\hat{J}_v$ are respectively the angular momentum operators for the boundary ($|\chi_b\rangle$) and the bulk state($|\psi_v\rangle$).

Local spacetime rotation, as a part of local Lorentz invariance, leaves quantum bulk Hilbert space invariant. Hence angular momentum operator, being the generator of spacetime rotation, annihilate the bulk states i.e.

$$\hat{J}_v|\psi_v\rangle = 0 \tag{9}$$

So Eqs. (5), (7) and (9) together imply,

$$\left[\widehat{H_v} - \beta\Phi\widehat{Q_v} - \beta\Omega\widehat{J_v}\right]|\psi_v\rangle = 0 \tag{10}$$

where, Φ, β and Ω are arbitrary functions at this stage.

2.2 Grand Canonical partition function

We will now consider a grand canonical ensemble of quantum spacetimes with horizons as boundaries, in contact with a heat bath, at some (inverse) temperature β. We will assume that this grand canonical ensemble of massive rotating charged black holes can exchange energy, angular momentum and electric charge with the heat bath. Therefore the grand canonical partition function becomes,

$$Z_G = Tr\left(exp\left(-\beta\hat{H} + \beta\Phi\hat{Q} + \beta\Omega\hat{J}\right)\right) \tag{11}$$

where the trace is taken over all states. Φ and Ω are respectively electrostatic potential and angular velocity of the black hole on the horizon.

Hence Eqs. (3), (4), (6), (8), (10) and (11) together yield

$$\begin{aligned} Z_G &= \sum_{b,v} |C_{b,v}|^2 \left\langle \psi_v \right| \otimes \left\langle \chi_b \right| exp\left(-\beta\hat{H} + \beta\Phi\hat{Q} + \beta\Omega\hat{J}\right) \left|\chi_b\right\rangle \otimes \left|\psi_v\right\rangle \\ &= \sum_b |C_b|^2 \left\langle \chi_b \right| exp\left(-\beta\hat{H}_b + \beta\Phi\hat{Q}_b + \beta\Omega\hat{J}_b\right) \left|\chi_b\right\rangle \end{aligned} \tag{12}$$

assuming that the boundary states can be normalized through the squared norm $\sum_v |c_{vb}|^2 \langle\psi_v|\psi_v\rangle = |C_b|^2$. This is analogous to the canonical ensemble scenario described in [14].

The partition function thus turns out to be completely determined by the boundary states (Z_{Gb}), i.e.,

$$Z_G = Z_{Gb} = Tr_b \exp\left(-\beta\hat{H}_b + \beta\Phi\hat{Q}_b + \beta\Omega\hat{J}_b\right) \tag{13}$$

The spectrum of the boundary Hamiltonian operator is assumed to be a function of the discrete electric charge and angular momentum spectrum associated with the horizon[1]. The total electric charge of a black hole is proportional to some fundamental charge from a quantum mechanical point of view and hence the electric charge spectrum is considered to be equispaced [16–20]. In fact the angular momentum spectrum can also be considered as equispaced in the macroscopic spectrum limit of the black hole [21], in which we are ultimately interested.

It has already been seen that electric charge, horizon area and angular momentum operators of a black hole commute among them and hence they can be diagonalized simultaneously. Therefore working in such diagonalized basis, the partition function (13) becomes

$$Z_G = \sum_{k,l,m} g(k,l,m) \quad \exp\left(-\beta(E(A_k, Q_l, J_m) - \Phi Q_l - \Omega J_m)\right) \tag{14}$$

where $g(k,l,m)$ is the degeneracy factor. k,l,m are respectively the quantum numbers corresponding to eigenvalues of horizon area, electric charge and angular momentum. In the macroscopic spectra limit of quantum isolated horizons i.e. regime of the large area, electric charge and angular momentum eigenvalues ($k \gg 1, l \gg 1, m \gg 1$), the Poisson resummation formula [22] implies

$$Z_G = \int dx\, dy\, dz\, g(A(x), Q(y), J(z)) \quad \exp\left(-\beta(E(A(x), Q(y), J(z)) - \Phi Q(y) - \Omega J(z))\right) \tag{15}$$

where x, y, z are respectively the continuum limit of k, l, m respectively.

---

[1] Actually this second assumption follows from [13, 15] for spacetimes admitting weakly isolated horizons where there exists a mass function determined by the area and electric charge associated with the horizon. This is an extension of that assumption to the quantum domain.

Now, A, Q and J are respectively, functions of x, y and z alone. Therefore we have,

$$dx = \frac{dA}{A_x}, dy = \frac{dQ}{Q_y}, dz = \frac{dJ}{J_z}$$

where, $A_x \equiv \frac{dA}{dx}$ and so on.

So, the partition function, in terms of area, electric charge and angular momentum as free variables, can be written as follows

$$Z_G = \int dA\, dQ\, dJ \quad \exp\left[S(A) - \beta(E(A,Q,J) - \Phi Q - \Omega J)\right], \tag{16}$$

where, following [23], the *microcanonical* entropy of the horizon is defined by $\exp S(A) \equiv \frac{g(A(x),Q(y),J(z))}{\frac{dA dQ dJ}{dx\, dy\, dz}}$ and is a function of horizon area(A) alone [10, 11, 24].

2.3 Stability against Gaussian fluctuations

2.3.1 Saddle point approximation

The equilibrium of a black hole is given by the saddle point $(\overline{A}, \overline{Q}, \overline{J})$ in the space of integration over horizon area, electric charge and angular momentum. It is now to study the grand canonical partition function for fluctuations $a = (A - \overline{A})$, $q = (Q - \overline{Q}), j = (J - \overline{J})$ around the saddle point to determine the stability of the black hole under Hawking radiation. We as usual restrict ourselves only up to Gaussian fluctuations, in order to extremize the free energy for the most probable configuration. Taylor expanding Eq. (16) about the saddle point, gives

$$\begin{aligned} Z_G = &\exp\left[S(\overline{A}) - \beta M(\overline{A},\overline{Q},\overline{J}) + \beta\Phi\overline{Q} + \beta\Omega\overline{J}\right] \\ &\times \int da\, dq\, dj \, \exp\left\{-\frac{\beta}{2}\left[\left(M_{AA} - \frac{S_{AA}}{\beta}\right)a^2 + (M_{QQ})q^2 + (2M_{AQ})aq \right.\right. \\ &\left.\left. + (M_{JJ})j^2 + (2M_{AJ})aj + (2M_{QJ})qj\right]\right\}, \end{aligned} \tag{17}$$

where $M(\overline{A},\overline{Q},\overline{J})$ is the mass of the isolated horizon at equilibrium. Here $M_{AQ} \equiv \left.\frac{\partial^2 M}{\partial A \partial Q}\right|_{(\overline{A},\overline{Q},\overline{J})}$ etc. and they are evaluated on the horizon. We will take the entropy of a black hole as linear in horizon area and hence S_{AA} equals to zero.

Now, in the saddle point approximation the coefficients of terms linear in a, q, j vanish by definition of the saddle point. These imply that

$$\beta = \frac{S_A}{M_A}, \Phi = M_Q, \Omega = M_J \tag{18}$$

Of course these derivatives are evaluated at the saddle point.

2.3.2 Criteria

Convergence of the integral (17) implies that the Hessian matrix (H) has to be positive definite, where

$$H = \begin{pmatrix} \beta M_{AA} & \beta M_{AQ} & \beta M_{AJ} \\ \beta M_{AQ} & \beta M_{QQ} & \beta M_{JQ} \\ \beta M_{AJ} & \beta M_{JQ} & \beta M_{JJ} \end{pmatrix} \tag{19}$$

The necessary and sufficient conditions for a real symmetric square matrix to be positive definite are: determinants of all principal square submatrices, and the determinant of the full matrix, are positive [25]. This condition leads to the following 'stability criteria':

$$M_{AA} > 0 \tag{20}$$

$$M_{QQ} > 0 \tag{21}$$

$$M_{JJ} > 0 \tag{22}$$

$$\left(M_{QQ}M_{JJ} - (M_{JQ})^2\right) > 0 \tag{23}$$

$$\left(M_{JJ}M_{AA} - (M_{AJ})^2\right) > 0 \tag{24}$$

$$\left(M_{QQ}M_{AA} - (M_{AQ})^2\right) > 0 \tag{25}$$

$$\begin{aligned}\Big[M_{AA}\left(M_{QQ}M_{JJ} - (M_{JQ})^2\right) - M_{AQ}(M_{AQ}M_{JJ} - M_{JQ}M_{AJ}) \\ + M_{AJ}(M_{AQ}M_{JQ} - M_{QQ}M_{AJ})\Big] > 0\end{aligned} \tag{26}$$

Of course, (inverse) temperature β is assumed to be positive for a stable configuration.

Now, the temperature is defined as, $T \equiv \frac{1}{\beta} = \frac{M_A}{S_A}$ (From Eq. (18)).

The relation $T = \frac{M_A}{S_A}$ implies that,

$$\frac{dT}{dA} = \frac{\beta M_A M_{AA}}{(S_A)^2} \tag{27}$$

Hence positivity of M_{AA} implies that a stable black hole becomes hotter as it grows in size. Schwarzschild black hole, violating this, invites its own thermal instability and decays under Hawking radiation [22].

It is obvious from Eq. (18) that, $M_{QQ} = \frac{d\Phi}{dQ}$ and $M_{JJ} = \frac{d\Omega}{dJ}$. Hence positivity of M_{QQ} implies that accumulation of charge increases the electric potential of the black hole, whereas positivity of M_{JJ} implies that accumulation of angular momentum makes the black hole to rotate faster. These are the features of a stable black hole (22).

The conditions for the convergence of grand partition function under Gaussian fluctuation imply the convexity of entropy [22, 23, 26]. Thus the above inequalities are correctly the conditions for thermal stability of a charged rotating black hole. Eqs. (20) and (27) together correctly reproduce that positivity of specific heat is the only criteria for thermal stability of an electrically neutral non rotating black hole [14]. Actually both mass and temperature of such black holes are functions of the horizon area (A) only and hence specific heat(C) is given as,

$$C \equiv \frac{dM}{dT} = \frac{(S_A)^2}{\beta M_{AA}} \tag{28}$$

Eqs. (20), (21) and (25) together describe the thermal stability of a non rotating electrically charged black hole, while (20), (22) and (24) together describe the same for rotating electrically neutral black holes [27]. Thus we find that positivity of specific heat cannot be the only criteria for thermal stability of an electrically charged rotating black hole, unlike Schwarzschild black hole, but the charge and the angular momentum play vital roles as well.

So far we have considered only the quantum version of a classical charged rotating black hole. But a quantum black hole may have other types of quantum charges as well. Hence we will consider all the charges of a quantum black hole in the same footing including angular momentum and electric charge. We consider a quantum black hole with n charges $C^1, \ldots, C^n$. Now following exactly the same prescription for constructing grand canonical partition function from operator algebra, we get the partition function here as,

$$
\begin{aligned}
Z_G = \exp&\left[S(\overline{A}) - \beta M\left(\overline{A}, \overline{C}^1, \ldots, \overline{C}^n\right) + \beta P_i \overline{C}^i\right] \\
&\times \int dA \left(\prod_{i=1}^{n} \int dC^i\right) \exp\left\{-\frac{1}{2}\left[\left(M_{AA}a^2 + 2\sum_{i=1}^{n} \beta M_{AC^i} a c^i\right.\right.\right. \\
&\left.\left.\left. + \sum_{i=1}^{n}\sum_{j=1}^{n} \beta M_{C^iC^j} c^i c^j\right)\right]\right\},
\end{aligned}
\tag{29}
$$

where $M\left(\overline{A}, \overline{C}^1, \ldots, \overline{C}^n\right)$ is the mass of equilibrium isolated horizon and $M_{AC^i} \equiv \partial^2 M / \partial A \partial C^i \big|_{\left(\overline{A}, \overline{C}^1, \ldots, \overline{C}^n\right)}$ etc., are evaluated on the horizon.

Convergence of the above integral (29) implies that the Hessian matrix (H) has to be positive definite, where

$$
H = \begin{pmatrix}
\beta M_{AA} & \beta M_{AC^1} & \beta M_{AC^2} & \ldots\ldots\ldots & \beta M_{AC^n} \\
\beta M_{AC^1} & \beta M_{C^1C^1} & \beta M_{C^1C^2} & \ldots\ldots\ldots & \beta M_{C^1C^n} \\
\beta M_{AC^2} & \beta M_{C^2C^1} & \beta M_{C^2C^2} & \ldots\ldots\ldots & \beta M_{C^2C^n} \\
.\ldots\ldots & .\ldots\ldots & \ldots\ldots & \ldots\ldots\ldots & .\ldots\ldots \\
\beta M_{AC^n} & \beta M_{C^nC^1} & \beta M_{C^nC^2} & \ldots\ldots\ldots & \beta M_{C^nC^n}
\end{pmatrix}
\tag{30}
$$

Here, all the derivatives are calculated at the saddle point. Hence the stability criteria i.e. the criteria for positive definiteness of Hessian matrix are given as:

$$
D_1 > 0, D_2 > 0, \ldots\ldots, D_{n+1} > 0 \tag{31}
$$

where,

$$
\begin{aligned}
&D_1 = \beta M_{AA}, \qquad D_2 = \begin{vmatrix} \beta M_{AA} & \beta M_{AC^1} \\ \beta M_{AC^1} & \beta M_{C^1C^1} \end{vmatrix}, \\
&D_3 = \begin{vmatrix} \beta M_{AA} & \beta M_{AC^1} & \beta M_{AC^2} \\ \beta M_{AC^1} & \beta M_{C^1C^1} & \beta M_{C^1C^2} \\ \beta M_{AC^2} & \beta M_{C^2C^1} & \beta M_{C^2C^2} \end{vmatrix}, \ldots\ldots, D_{n+1} = |H|
\end{aligned}
\tag{32}
$$

where, $|H|$ = determinant of the Hessian matrix H.

The inverse temperature β is expectantly assumed to be positive for a stable black hole. We again find that temperature must increase with horizon area, inherent in the positivity of M_{AA}. $'n'$ equals two for a charged rotating black hole and hence according to (32) there should be three stability criteria, not seven (20)–(26). It is to note that those seven conditions are not all independent, actually only three of them are independent.

3. Quasi stability, thermal fluctuations and phase transitions of black holes

Some black holes may not satisfy all the stability criteria together everywhere in their parameter spaces. Such regimes are regions of quasi stability for that black hole and the black hole is quasi stable in that regime. Thus quasi stability of a black hole depends entirely on the regime of parameter space where the black hole is. Of course certain stability criteria may not hold anywhere in parameter space for some black holes and they are completely quasi stable. We will see the relationship between quasi stability and thermal fluctuation in this section.

We found for stable black holes that the grand canonical partition function is converging. We can hence define fluctuation of their parameters. The standard deviation of the statistical distribution of a quantity measures the expectation value of its fluctuation. This knowledge along with the grand canonical partition function implies the standard deviation of charge(Q) as,

$$(\Delta Q)^2 = \frac{\int da\, dq\, dj\, q^2 \exp\left\{-\frac{\beta}{2}\left[\left(M_{AA} - \frac{S_{AA}}{\beta}\right)a^2 + (M_{QQ})q^2 + (2M_{AQ})aq + (M_{JJ})j^2 + (2M_{AJ})aj + (2M_{QJ})qj\right]\right\}}{\int da\, dq\, dj\, \exp\left\{-\frac{\beta}{2}\left[\left(M_{AA} - \frac{S_{AA}}{\beta}\right)a^2 + (M_{QQ})q^2 + (2M_{AQ})aq + (M_{JJ})j^2 + (2M_{AJ})aj + (2M_{QJ})qj\right]\right\}} \tag{33}$$

where, ΔQ is the standard deviation for the electric charge of the black hole. Similarly, ΔJ and ΔA are respectively the same for angular momentum and horizon area of the black hole.

Both the numerator and denominator are converging and turns out to be,

$$(\Delta Q)^2 = -\frac{2}{\beta} \cdot \frac{1}{Z_G} \cdot \frac{\partial Z_G}{\partial M_{QQ}} = \frac{1}{|H|} \cdot \frac{\partial |H|}{\partial (\beta M_{QQ})} \tag{34}$$

where, $|H|$ = determinant of Hessian matrix(H).

The above said process is invalid for quasi stable black holes as their grand canonical partition functions diverge. Hence necessary rearrangements are required to express their grand canonical partition function in the diagonal basis of their Hessian matrices and then to look for stable modes. Fortunately fluctuations of these stable modes are calculable and finite, although their grand partition functions diverge.

We can now rewrite the grand canonical partition function (Z_G) in the diagonal basis of the Hessian matrix as,

$$Z_G = \left(\prod_{j=1}^{n+1} \int d\underline{c}^j\right) \exp\left\{-\frac{1}{2}\left[D_1 \left(\underline{c}^1\right)^2 + \frac{D_2}{D_1}\left(\underline{c}^2\right)^2 + \ldots + \frac{D_{n+1}}{D_n}\left(\underline{c}^{n+1}\right)^2\right]\right\} \tag{35}$$

where the expressions of $D_1, D_2, , D_{n+1}$ are the same as given in (20). The new variables $(\underline{c}^1, , \underline{c}^{n+1})$ are related to the old variables $(a, c^1, , c^n)$ by some linear transformation. The linear transformation matrix is a $(n+1)$ dimensional upper triangular square matrix and hence it has unit determinant. The elements of this transformation matrix are functions of the elements of the Hessian matrix H. Thus it is obvious that exactly one of the $\underline{c}^j$ is equal to C^j, but that identification is not unique. This actually helps us to calculate the fluctuation of any parameter of quasi stable black hole that we want. If at least one of $D_1, \frac{D_2}{D_1}, \dots, \frac{D_{n+1}}{D_n}$ is negative, then Z_G blows up.

We can now define fluctuations for quasi stable black holes in the same way as we did for stable black holes. If D_1 is positive then the fluctuation of $\underline{c}^1$ $\left(\Delta(\underline{c}^1)^2\right)$ is finite and equals to $\frac{1}{2D_1}$, otherwise it blows up. Similarly $\Delta(\underline{c}^2)^2, \Delta(\underline{c}^3)^2, \dots, \Delta(\underline{c}^{n+1})^2$ can be defined and equal to $\frac{D_1}{2D_2}, \frac{D_2}{2D_3}, \dots, \frac{D_n}{2D_{n+1}}$ respectively only if these ratios of the coefficients are positive.

A stable black hole with n charges possesses $(n+1)$ independent thermal stability conditions [5]. But it was already shown that an electrically charged, rotating stable black hole possessed seven conditions for thermal stability [4]. But only three of them are independent, the rest depend on those three conditions. But this conclusion holds only for stable black holes, not for quasi stable black holes. Thus one has to check the positivity of determinants of all $(2^{n+1}-1)$ submatrices of Hessian matrix H (including itself) to ensure the quasi stability of a black hole.

Thus we see that stability of a black hole is determined by the signs of the functions, appeared in the stability criteria. There will be $(n+1)$ no. of fluctuations for a black hole having $'n'$ no. of charges. These fluctuations are individually related, to be shown later, with some physical quantities of the black hole. Signs of each of these physical quantities designate one distinguished phase. Thus a quasi stable black hole with $'n'$ charges can at most have 2^{n+1} number of phases. Any of these physical quantities can possess the same sign in different regimes of parameter space and hence the black hole can enter in the same phase once again. So a decaying black hole may be lucky enough to enjoy the phases of its younger age once more. These interesting reoccurrence of phase transitions are completely absent in both stable or unstable black holes. The relationship among the boundary degrees of freedom determines these phases in a quasi stable black hole.

Finite, bounded fluctuations of the parameters of both stable and quasi stable black holes are directly connected with their respective stability criteria [8, 9]. These fluctuations will be shown to be related with some physically measurable quantities of the black hole. Flipping of their signs indicate phase transitions, generalization of Hawking's old idea for asymptotically flat Schwarzschild black hole (AFSBH) [6] but in case of quasi stable black holes. Hawking showed that negative specific heat made AFSBH thermally unstable. Divergence in ΔA^2 made it happen for AFSBH [8]. But quasi stable black holes possess too many parameters, other than horizon area. Hence fluctuations of other parameters are similarly expected to be related with other physical quantities of the black hole. We will see soon that this expectation is actually the reality.

We will now use the summation formalism of partition function to build up various physical quantities in connection with quasi stable black holes.

In this formalism, grand canonical partition function is given as [4],
$Z_G = \sum_r \exp\left(-\beta(E_r - \Phi Q_r - \Omega J_r)\right)$; here summation is taken over eigenstates.
The various symbolic terms like Φ, Ω etc. are as before.

Define, $\overline{\Phi} \equiv \beta\Phi$ and $\overline{\Omega} \equiv \beta\Omega$. $\overline{\Phi}$ and $\overline{\Omega}$ respectively determines the electrical and rotational equilibrium between two connected systems [28].

Hence the grand canonical partition function becomes,

$$Z_G = \sum_r \exp\left(-\beta E_r + \overline{\Phi} Q_r + \overline{\Omega} J_r\right).$$

Thus, average value of angular momentum can be defined as,

$$\bar{J} \equiv \frac{\sum_r J_r \cdot \exp\left(-\beta E_r + \overline{\Phi} Q_r + \overline{\Omega} J_r\right)}{Z_G} = \partial(\ln(Z_G))/\partial\overline{\Omega}$$

Similarly we can calculate $\bar{J}^2$ and is given as,

$$\bar{J}^2 \equiv \frac{\sum_r J_r^2 \cdot \exp\left(-\beta E_r + \overline{\Phi} Q_r + \overline{\Omega} J_r\right)}{Z_G}$$

We can calculate fluctuation of angular momentum and this turns out to be

$$\Delta(J)^2 \equiv \frac{\sum_r (J_r - \bar{J})^2 \cdot \exp\left(-\beta E_r + \overline{\Phi} Q_r + \overline{\Omega} J_r\right)}{Z_G} = \bar{J}^2 - (\bar{J})^2 = \partial^2(\ln(Z_G))/\partial\overline{\Omega}^2$$

The convergence of fluctuation for angular momentum is mandatory for the above calculation. Most importantly the above partial derivatives are taken at the constant values of β and $\overline{\Phi}$. Likewise partial derivatives with respect to $\overline{\Phi}$ can be taken at constant values of β, $\overline{\Omega}$ and so on.

The rotational inertia of a black hole (S_J) is defined as,

$S_J \equiv \beta \cdot \partial\bar{J}/\partial\overline{\Omega}$ and is equals to $\beta \cdot \Delta(J)^2$.

It is important to note the following issue:

The quantities β, $\overline{\Phi}$ and $\overline{\Omega}$ are functions of independent variables $\overline{A}$, $\overline{Q}$ and $\bar{J}$ and hence consequently $\overline{A}$, $\overline{Q}$ and $\bar{J}$ are the inverse functions of β, $\overline{\Phi}$ and $\overline{\Omega}$. Hence partial derivatives for example with respect to $\overline{\Omega}$, at constant β, $\overline{\Phi}$, can be evaluated and so on. So S_J and $\Delta(J)^2$ are independently calculable. They are related only when fluctuation in angular momentum is bounded and finite. $\Delta(J)^2$ approaches zero and then suddenly blows up at the point of phase transition. But S_J vanishes there and flips its sign afterwards. It starts to disrespect the above equality afterwards.

Electric capacitance of a black hole(S_Q) is defined as,

$S_Q \equiv \beta \cdot \partial\overline{Q}/\partial\overline{\Phi}$ and is equal to $\beta \cdot \Delta(Q)^2$, only when $\Delta(Q)^2$ is finite and bounded.

S_Q and $\Delta(Q)^2$ respectively are in same footings as that of S_J and $\Delta(J)^2$ regarding their relationship and behavior at the point of phase transition. Hence flipping in signs of electric capacitance and rotational inertia separately mark two different phase transitions.

4. Decay of quasi stable black holes and possible identification with dark matter

4.1 Asymptotically flat Reissner-Nordstrom black hole

The mass(M) of asymptotically flat Reissner-Nordstrom black hole (AFRNBH) depends on its parameters as [29],

$$M = \frac{\sqrt{A}}{4\sqrt{\pi}} + \frac{\sqrt{\pi} Q^2}{\sqrt{A}} \tag{36}$$

We can now calculate the temperature of AFRNBH and it will be function of its electric charge(Q) and horizon area(A). On calculation, it turns out that temperature $(\propto M_A)$ is positive only if $\frac{Q^2}{A} < \frac{1}{4\pi}$. This restricts the parameter space.

We can calculate various second derivatives of the black hole mass (M) with respect to its parameters from the above relation. On calculation, this turns out that.

$$M_{QQ} = \frac{2\sqrt{\pi}}{\sqrt{A}}, M_{AQ} = -\frac{\sqrt{\pi}Q}{A^{3/2}}, M_{AA} = -\frac{1}{16\sqrt{\pi}A^{3/2}} + \frac{3\sqrt{\pi}Q^2}{4A^{5/2}}, \left(M_{QQ}M_{AA} - (M_{AQ})^2\right) = \left(-\frac{1}{8A^2} + \frac{\pi Q^2}{2A^3}\right)$$

Thus $\left(M_{QQ}M_{AA} - (M_{AQ})^2\right)$ is positive only if $\frac{Q^2}{A} > \frac{1}{4\pi}$. But this region of parameter space is not accessible to any real AFRNBH as it is excluded due to negativity of temperature. Hence $\left(M_{QQ}M_{AA} - (M_{AQ})^2\right)$ is negative throughout its physically accessible regime of parameter space. Now, M_{QQ} is always positive while M_{AA} is negative if $\frac{Q^2}{A} < \frac{1}{12\pi}$. Thus AFRNBH can never be thermally stable as it never satisfies any of the above two stability criteria completely. So AFRNBH is actually a quasi stable black hole [9].

Now, $\left(M_{QQ}M_{AA} - (M_{AQ})^2\right)$ is always negative for AFRNBH. Keeping this in mind, We can conclude that,

1. $\Delta(A)^2$ always blows up as M_{QQ} is always positive.

2. $\Delta(Q)^2$ converges and equals to the $\frac{M_{AA}}{2\beta\left(M_{QQ}M_{AA} - (M_{AQ})^2\right)}$ only if $M_{AA} < 0$ i.e. $\frac{Q^2}{A} < \frac{1}{12\pi}$.

AFRNBH gradually becomes smaller in size due to unbounded area fluctuation and hence ultimately decays. Thus $\frac{Q^2}{A}$, even if it is less than $\frac{1}{12\pi}$ at the beginning, increases as area(A) decreases. But it cannot go beyond $\frac{1}{4\pi}$. In the regime $\frac{1}{4\pi} > \frac{Q^2}{A} > \frac{1}{12\pi}$, electric charge$(Q)$ of this black hole fluctuates appreciably enough to reduce the value of Q. Thus this ratio becomes lower than the bench mark value $\frac{1}{12\pi}$. Hence we see that this toggling keeps on going around the value $\frac{1}{12\pi}$. In this process the black hole will continue to lose its electric charge and horizon area and consequently moves forward to its end state with a certain minimum area [30], having almost no electric charge. At this point, the black hole will not decay any further and becomes thermodynamically isolated. Only gravitational interaction remains active. This is quite similar to the nature of dark matter. This correspondence is possible only if we are ready to accept that what we think of as dark matter is actually some region of the spacetime of our universe. Thus this region pretends to be neutral Planck dark matter as the size of black hole is now of the order of Planck length.

4.2 Asymptotically flat Kerr-Newman black hole

The mass(M) of this black hole depends on its parameters as [31],

$$M^2 = \frac{A}{16\pi} + \frac{\pi}{A}(4J^2 + Q^4) + \frac{Q^2}{2}$$

So the parameter space is restricted by the inequality $(4J^2 + Q^4) < \frac{A^2}{16\pi^2}$ as temperature($\propto M_A$) of a non extremal black hole is always positive. Hence both electric charge and angular momentum are bounded for a given horizon area of the black hole. $|H|$ can be shown to be always negative and hence this black hole would decay under Hawking radiation. It will consequently lose its area. Hence charge and angular momentum have to adjust them respectively through their fluctuations to maintain the above bound. This bounded region is shown in the **Figure 1**.

Now, it can be easily shown that $\left(M_{QQ}(\beta M_{AA} - S_{AA}) - \beta(M_{AQ})^2\right)$ is negative in the upper portion of the shaded region of the above figure. Thus this is the region for bounded fluctuation of angular momentum. So, the higher values of $\frac{J}{A}$ make the fluctuation of angular momentum large. As the area of this black hole always decreases, the ratio $\frac{J}{A}$ increases. Thus the fluctuation of angular momentum becomes appreciably large and hence angular momentum is reduced to maintain the non extremality bound. So, $\frac{J}{A}$ ratio again comes to the regime where J does not fluctuate much. But area(A) as usual decreases continuously and consequently $\frac{J}{A}$ ratio again becomes large enough such that J starts to fluctuate appreciably again. Thus this flipping of $\frac{J}{A}$ ratio from larger to smaller value and vice versa keeps on going. Hence angular momentum gradually decreases and consequently KN black hole proceeds to transform into a non rotating black hole.

On the other hand, it can be easily shown that $\left(M_{JJ}(\beta M_{AA} - S_{AA}) - \beta(M_{AJ})^2\right)$ is negative in the lower portion of the shaded region of the above figure. Thus this is the region for bounded fluctuation of charge. So, higher values of $\frac{Q^2}{A}$ make the fluctuation of charge bounded only if the ratio $\frac{J}{A}$ is sufficiently high. But we have just seen that $\frac{J}{A}$ ratio cannot always be high, along with the fact that the ratio $\frac{Q^2}{A}$ is itself bounded. Thus Q reduces gradually as the area of the black hole decreases. So, the ratio $\frac{Q^2}{A}$ oscillates between higher and lower values, exactly in the same manner as $\frac{J}{A}$ ratio does the same and gradually discharges all its charges. Consequently it proceeds to transform into a chargeless, non rotating black hole. Thus it resembles neutral Planck dark matter due to the fact explained in the last section. The difference between this sort of dark matter and the earlier one is only that their origins are different.

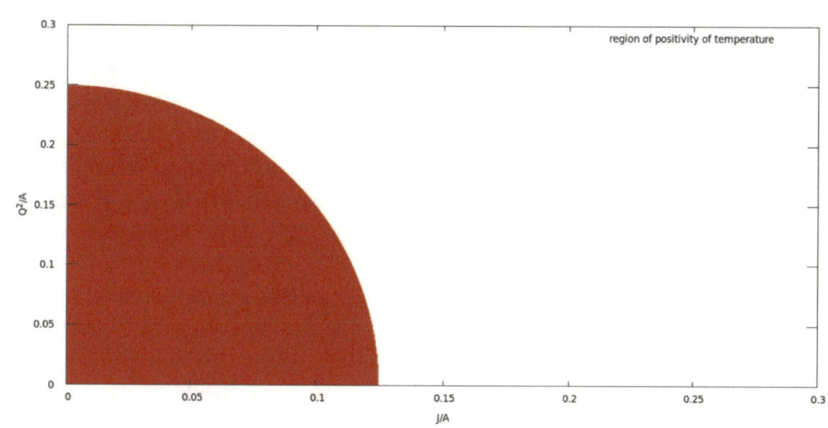

Figure 1.
Pictorial representation of region of positivity of temperature.

4.3 Asymptotically flat Kerr-Sen black hole

The mass(M) of this black hole depends on its parameters as [32],

$$M^2 = \frac{A}{16\pi} + \frac{Q^2}{2} + \frac{4\pi J^2}{A}$$

The parameter space here is restricted by the inequality $\frac{J}{A} < \frac{1}{8\pi}$ as temperature ($\propto M_A$) of a non extremal black hole is always positive. It is important to notice that the electric charge of this black hole, unlike AFKNBH, is not bounded by the non extremality of this black hole. We will see its interesting consequences soon. The quantity $\left(M_{QQ}(\beta M_{AA} - S_{AA}) - \beta(M_{AQ})^2\right)$ is negative in the regime $\frac{J}{A} < \frac{0.4}{8\pi}$, but $\left((\beta M_{AA} - S_{AA})M_{JJ} - \beta(M_{JA})^2\right)$ is always negative. Hence both $\Delta(J)^2$ and $\Delta(Q)^2$ are bounded in the regime $\frac{J}{A} < \frac{0.4}{8\pi}$ for KS black Hole, maintaining a perfect balance between the incoming and outgoing quanta of angular momentum and electric charge respectively. But this balance is lost only for angular momentum in the regime $\frac{0.4}{8\pi} < \frac{J}{A} < \frac{1}{8\pi}$, whereas the same for electric charge is maintained everywhere in the parameter space. But the KS black hole ultimately decays due to unbounded nature of $\Delta(A)^2$.

Suppose the angular momentum(J) is such that $\frac{J}{A} < \frac{0.4}{8\pi}$ and hence J does not fluctuate much as its fluctuation is bounded in this region. But area(A) as usual decreases and hence the ratio $\frac{J}{A}$ increases and becomes greater than $\frac{0.4}{8\pi}$. Once this ratio crosses that value, J starts to fluctuate rapidly. But this ratio, due to non extremality, cannot be greater than $\frac{1}{8\pi}$ with decreasing area(A). Thus J ultimately reduces and hence the ratio $\frac{J}{A}$ becomes lesser than $\frac{0.4}{8\pi}$. This process will go on. This means that KS black hole tries to reduce the angular momentum, in order to satisfy its extremality bound, during the Hawking decay. Hence the black hole gradually loses its area and angular momentum, keeping the charge unchanged. Thus it proceeds to transform into a black hole with charge only. This transformation is purely thermodynamical in nature. Thus we find the difference between KS and KN black hole in terms of their end states.

It is important to note that KS black hole, unlike KN black hole, hardly discharges throughout its life. One has to go back to the construction of grand canonical partition to understand this. We in this analysis have assumed the mass of a rotating charged black hole as a function of its area, charge and angular momentum. It is a fact in any theory of quantum gravity that area, charge and angular momentum are good self-adjoint operators. But mass is not a good primary operator. We still can represent it as a secondary operator in terms of other primary operators. Hence we here consider fluctuations of area, charge and angular momentum only. In semiclassical analyses, one gets various restrictions on the parameter space from the condition of avoiding the naked singularity. We, in thermodynamical analysis, equivalently obtain various restrictions on the parameter space from the condition of avoiding the absolute zero temperature. Semi classically, it had been shown [33] that a charged rotating KN black hole should lose its charge and angular momentum, just from the condition of various restrictions on the parameter space. We also obtain similar results for KN black holes, just from the condition of various restrictions on the parameter space imposed by positivity of the temperature. But this analysis is a bit interesting for KS black hole. Positivity of temperature does not put any bound on its electric charge. Close to the end state, this black hole loses almost all its angular momentum. The area also becomes comparable with the Planck area

[30]. Hence mass of the black hole is approximated given there as, $M^2 \approx \frac{Q^2}{2}$. This is very much similar to stable extremal black holes with magnetic monopoles. Of course the last example is the outcome of semiclassical analysis, where the mass of this black hole in the limiting case is given as, $M^2 \approx P^2$, P is magnetic charge. We compare this thermodynamical analysis with well known semiclassical analyses not to establish our analysis, but to show the simplicity as well as superiority of this analysis. B.carter, through his semi classical analysis [34], had shown that charged black hole with initial mass of order of 10^{15} kg does negligibly discharge throughout its life. This, if translated for KS black hole, implies KS black hole almost does not discharge if its initial charge is roughly one mole of electrons. In fact charged black holes with sufficient initial mass, under certain idealized conditions, had been shown semi classically [35] not to discharge. This again supports our conclusion regarding stability of electric charge for decaying KS black hole.

The end state of this black hole can now be identified as charged Planck dark matter. Thus we get a possible scenario for obtaining a charged black hole through our line of thoughts.

4.4 (2 + 1) dimensional charged BTZ black hole

The mass(M) of $(2+1)$ dimensional charged BTZ black hole (Λ3BTZBH) depends on its parameters as [36],

$$M = \frac{r^2}{8l^2} - \frac{Q^2}{16}\, ln\,(r/l).$$

Here l is known as cosmic length and is related with Λ as $\Lambda = 1/l^2$. r is the radius of the circular horizon. Hence area of it, which is actually its perimeter, is given as $A = 2\pi r$. So the mass(M) of Λ3BTZBH can be expressed in terms of Λ and A as,

$$M = \frac{A^2\Lambda}{32\pi^2} - \frac{Q^2}{32}\, ln\left(\frac{A^2\Lambda}{4\pi^2}\right) \tag{37}$$

We can now calculate the temperature of Λ3BTZBH from above relationship and it becomes a function of its charge(Q), area(A) and Λ. On calculation, it turns out that temperature($= M_A$) is positive if $A^2 > \pi^2 Q^2/\Lambda$. This restricts the parameter space.

We can calculate various second order derivatives of the black hole mass(M) with respect to its parameters from the above relationship. On calculation, this turns out that.

$$M_{QQ} = -\frac{1}{16}\, ln\left(\frac{A^2\Lambda}{4\pi^2}\right), M_{AQ} = -\frac{Q}{8A}, M_{Q\Lambda} = -\frac{Q}{16\Lambda}, M_{\Lambda\Lambda} = \frac{Q^2}{32\Lambda^2}, M_{\Lambda A} = \frac{A}{16\pi^2}, M_{AA} = \left(\frac{\Lambda}{16\pi^2} + \frac{Q^2}{16A^2}\right)$$

We, with the help of the above six second order derivatives of M, can show that.

1. $\left(M_{\Lambda\Lambda}M_{AA} - (M_{A\Lambda})^2\right) = \frac{A^2}{256\pi^4}\left(\frac{\pi^2 Q^2}{2\Lambda A^2} + \frac{\pi^4 Q^4}{2\Lambda^2 A^4} - \pi^2\right)$. Now positivity of temperature implies $A^2 > \pi^2 Q^2/\Lambda$. Hence $\left(M_{\Lambda\Lambda}M_{AA} - (M_{A\Lambda})^2\right)$ is always negative.

2. $|H| = ln\left(\frac{A^2\Lambda}{4\pi^2}\right) \cdot \frac{A^2}{256 \cdot 32\pi^4} \cdot \left(2 - \frac{\pi^2 Q^2}{2\Lambda A^2} - \frac{\pi^4 Q^4}{2\Lambda^2 A^4}\right) + \frac{6Q^2}{265 \cdot 32\pi^2\Lambda}\left(1 - \frac{\pi^2 Q^2}{A^2\Lambda}\right)$. Now again the positivity of temperature fixes the sign of $|H|$, but here it is positive.

Thus we explicitly see that $\left(M_{\Lambda\Lambda}M_{AA} - (M_{A\Lambda})^2\right)$ is always negative for Λ3BTZBH, whereas $|H|$ is always positive for it. Hence Λ3BTZBH is actually quasi stable under Hawking radiation. The most interesting point to note that $|H|$ is always positive here, unlike other quasi stable black holes [9, 28, 37].

We have earlier shown that [9, 37] quasi stable black holes possess tiny fluctuations for some of their parameters in certain regions of parameter space. So, the same is expected in case of Λ3BTZBH. We already knew [9] how fluctuations were related to stability criteria. In fact we also knew [9] how to calculate fluctuations in case of quasi stable black holes. Now, $|H|$ is always positive. Keeping this in mind, we can conclude[2] that,

1. $\Delta(A)^2$ is bounded only if $\left(M_{QQ}M_{\Lambda\Lambda} - (M_{\Lambda Q})^2\right)$ is positive. On calculation it turns out that, $\left(M_{QQ}M_{\Lambda\Lambda} - (M_{\Lambda Q})^2\right) = -\frac{Q^2}{256\Lambda^2}\left(1 + \frac{1}{2} \cdot ln\left(\frac{A^2\Lambda}{4\pi^2}\right)\right)$ and is positive if $A^2\Lambda < \frac{4\pi^2}{e^2}$.

2. $\Delta(Q)^2$ is always unbounded as $\left(M_{\Lambda\Lambda}M_{AA} - (M_{A\Lambda})^2\right)$, has already been shown, is always negative.

3. $\Delta(\Lambda)^2$ is bounded only if $\left(M_{QQ}M_{AA} - (M_{AQ})^2\right)$ is positive. On calculation it turns out that, $\left(M_{QQ}M_{AA} - (M_{AQ})^2\right) = -\frac{1}{16}\left(ln\left(\frac{A^2\Lambda}{4\pi^2}\right)\left(\frac{\Lambda}{16\pi^2} + \frac{Q^2}{16A^2}\right) + \frac{Q^2}{4A^2}\right)$ and is positive if $ln\left(\frac{A^2\Lambda}{4\pi^2}\right) < -\frac{4}{1 + \frac{A\Lambda}{\pi^2 Q^2}}$. Now positivity of temperature gives $A^2 > \pi^2 Q^2/\Lambda$ and consequently this implies $\left(-\frac{4}{1 + \frac{A\Lambda}{\pi^2 Q^2}}\right)$ is greater than -2. Thus $A^2\Lambda < \frac{4\pi^2}{e^2}$ is the region for positivity of $\left(M_{QQ}M_{AA} - (M_{AQ})^2\right)$. In fact this upper limit is greater than the estimated value as $\frac{A^2\Lambda}{\pi^2 Q^2}$ is greater than unity. So, $\pi^2 Q^2 < A^2\Lambda < \frac{4\pi^2}{e^2}$ is a legitimate regime in parameter space, where $\Delta\Lambda^2$ is bounded.

We have just seen that charge always fluctuates with large magnitude. Now, suppose area(A) is initially so large that it satisfies both the inequalities $\pi^2 Q^2 < A^2\Lambda$ and $A^2\Lambda > \frac{4\pi^2}{e^2}$ by far. In this regime of parameter space all the parameters charge (Q), area (A) and cosmological constant(Λ) together fluctuate appreciably. Area gradually decreases due to Hawking radiation. Cosmological constant also gradually decreases due to bubble emission [38]. Hence charge has to decrease sufficiently fast to maintain the positivity of temperature, as otherwise zero temperature would

---

[2] $\Delta(\Lambda)^2$ measures the fluctuation of cosmological constant from its equilibrium value and is mathematically expressed as [8, 9], $\Delta(\Lambda)^2 = \frac{\int da\, dq\, d\lambda\, \lambda^2 f(a,\lambda)}{\int da\, dq\, d\lambda}$, where $f(a,\lambda) = exp\left(-\frac{\beta}{2}\left[\left(M_{AA} - \frac{S_{AA}}{\beta}\right)a^2 + (M_{QQ})q^2 + (M_{\Lambda\Lambda})\lambda^2 + (2M_{AQ})aq + (2M_{A\Lambda})a\lambda + (2M_{Q\Lambda})q\lambda\right]\right)$. Similarly, $\Delta(A)^2$ and $\Delta(Q)^2$ are defined.

cause the thermodynamic death of the black hole. Thus the black hole would once cross the curve $A^2\Lambda = \frac{4\pi^2}{e^2}$, making the term $\left(M_{QQ}M_{\Lambda\Lambda} - (M_{\Lambda Q})^2\right)$ positive. Hence $\Delta(A)^2$ becomes bounded, suppressing its large unbounded magnitude exponentially. In fact $\left(M_{QQ}M_{AA} - (M_{AQ})^2\right)$ becomes positive even before $A^2\Lambda$ becomes equal to $\frac{4\pi^2}{e^2}$. This consequently makes $\Delta(\Lambda)^2$ bounded, like $\Delta(A)^2$, suppressing its large unbounded magnitude exponentially. Thus we find that in the regime $\pi^2Q^2 < A^2\Lambda < \frac{4\pi^2}{e^2}$, both area and cosmological constant do not fluctuate appreciably. But charge gradually decreases as before and hence the last inequality holds good. Thus once the black hole loses almost all its charge, it transforms into a stable chargeless BTZ black hole, having negative cosmological constant. This end state of Λ3BTZBH, as we have seen, is different from other AdS black holes as their horizon areas become close to the Planck area in their end states [39]. Thus we now get a non Planck sized dark matter from our line of thoughts.

5. Note

The readers may wonder how this chapter can be something about dark matter? We have hardly used the word "dark matter" so far, at most have used it on a few occasions. But theoretically the connection, which is discussed here, between black holes and dark matter is extremely appealing. There are some experimental evidences that mostly rule out the possibility of connection between dark matter and black hole [40], that we have described. On the other hand the recent observations of LIGO and VIRGO now suggest that black holes are much more common than once imagined and hence they could very well be the missing dark matter [41]. Anyway this chapter is written with the belief that dark matter is the possible end state of the quasi stable black holes. Many more future experiments are required to conclude definitively about the validity of our predictions.

Classification

PACS numbers: 04.70.-s, 04.70.Dy.

Author details

Aloke Kumar Sinha
Haldia Government College, West Bengal, India

*Address all correspondence to: akshooghly@gmail.com

References

[1] Weinberg S 1972 *Gravitation and Cosmology* ISBN: 978-0-471-92567-5

[2] Wald R M 1984 *General Relativity* ISBN: 978-0-226-87033-5

[3] Gibbons G W and Hawking S W 1977 Phys. Rev. D 15 273851; DOI: 10.1103/PhysRevD.15.2738

[4] Sinha A K and Majumdar P 2017 Mod. Phys. Lett. A 32 1750208; DOI: 10.1142/S021773231750208X

[5] Sinha A K 2018 Mod. Phys. Lett. A 33 1850031; DOI: 10.1142/S0217732318500311

[6] Hawking S W and Page D N 1983 Commun. Math. Phys. 87 577; DOI: 10.1007/BF01208266

[7] Davis P C W 1977 Proc. Roy. Soc. A 353; DOI: 10.1098/rspa.1977.0047 499

[8] Sinha A K 2018 Mod. Phys. Lett. A 33 1850190; DOI: 10.1142/S0217732318501900

[9] Sinha A K 2019 Class. Quant. Grav. 36 035003; DOI: 10.1088/1361-6382/aafa57

[10] Ashtekar A, Baez J, Corichi A and Krasnov K 1998 Phys. Rev. Lett. 80 904; DOI: 10.1103/PhysRevLett.80.904

[11] Ashtekar A, Baez J and Krasnov K 2000 Adv. Theor. Math. Phys. 4 1; DOI: 10.4310/ATMP.2000.v4.n1.a1

[12] Thiemann T 2007 *Modern Canonical Quantum General Relativity* ISBN: 9780511755682

[13] Ashtekar A, Fairhurst S and Krishnan B 2000 Phys. Rev. D 62 104025; DOI: 10.1103/PhysRevD.62.104025

[14] Majumdar P 2007 Class. Quant. Grav. 24 1747; DOI: 10.1088/0264-9381/24/7/005

[15] Ashtekar A and Krishnan B 2004 Living Rev. Rel. 7 10; DOI: 10.12942/lrr-2004-10

[16] Ashtekar A, Beetle C and Lewandowski J 2001 Phys. Rev. D 64 044016; DOI: 10.1103/PhysRevD.64.044016

[17] Majhi A and Majumdar P 2014 Class. Quant. Grav. 31 195003; DOI: 10.1088/0264-9381/31/19/195003

[18] Kaul R K and Rama S K 2003 Phys. Rev. D 68 024001; DOI: 10.1103/PhysRevD.68.024001

[19] Frodden E, Perez A, Pranzetti D and Roken C 2014 Gen Relativ Gravit 46 1828; DOI: 10.1007/s10714-014-1828-6

[20] Achour J B, Noui K and Perez A 2016 JHEP 1608 149

[21] Gour G and Medved A J M 2003 Class.Quant.Grav. 20 2261-2274; DOI: 10.1088/0264-9381/20/11/321

[22] Chatterjee A and Majumdar P 2004 Phys. Rev. Lett. 92 141031; DOI: 10.1103/PhysRevLett.92.141301

[23] Landau L D and Lifschitz E M 1980 *Statistical Physics* ISBN: 978-0-08-057046-4

[24] Kaul R K and Majumdar P 2000 Phys. Rev. Lett 84 5255; 10.1103/PhysRevLett.84.5255

[25] Meyer C D 2000 *Matrix Analysis and Applied Linear Algebra* ISBN: 0-89871-454-0; Bhatia R 2007 *Positive Definite Matrices* ISBN: 9780691168258; Pinkus A 2009 *Totally Positive Matrices* ISBN: 9780521194082

[26] Monteiro R 2010, e-Print arxiv:hep-th 1006.5358

[27] Majhi A and Majumdar P 2012 Class. Quant. Grav. 29 135013; DOI: 10.1088/0264-9381/29/13/135013

[28] Sinha A K 2020 Mod. Phys. Lett. A 35 2050258; DOI: 10.1142/S0217732320502582

[29] Carter B 1966 Physics Letters 21 423-424; DOI: 10.1016/0031-9163(66)90515-4

[30] Rovelli C and Vidotto F 2018 Universe 4 127; DOI: 10.3390/universe4110127

[31] Caldarelli M M, Cognola G and Klemm D 2000 Class. Quant. Grav. 17 399; DOI: 10.1088/0264-9381/17/2/310

[32] Sen A 1992 Phys. Rev. Lett. 69 1006; DOI: 10.1103/PhysRevLett.69.1006

[33] Page D N 1976 Phys. Rev. D 14 3260; DOI: 10.1103/PhysRevD.14.3260

[34] Carter B 1974 Phys. Rev. Lett. 33 558; DOI: 10.1103/PhysRevLett.33.558

[35] Hiscock W A and Weems L D 1990 Phys. Rev. D 41 1142; DOI: 10.1103/PhysRevD.41.1142

[36] Frassino A M, Mann R B and Mureika J R 2015 Phys. Rev. D 92 124069; DOI: 10.1103/PhysRevD.92.124069

[37] Sinha A K 2020 Mod. Phys. Lett. A 35 2050136; DOI: 10.1142/S0217732320501369

[38] Teittelboim C 1985 Phy. letters 158B 4; DOI: 10.1016/0370-2693(85)91186-4

[39] Sinha A K 2021 Mod. Phys. Lett. A 36 2150071; DOI: 10.1142/S0217732321500711

[40] Zumalacrregui M and Seljak U 2018 Phys. Rev. Lett. 121, 141101; DOI: 10.1103/PhysRevLett.121.141101

[41] https://pnp.ligo.org/ppcomm/Papers.html. This is the link for digital collections of publications of the LIGO scientific collaboration and VIRGO collaboration. I like to thank the editor for bringing this fact in my memory.

Chapter 6

Non-Keplerian Orbits in Dark Matter

Peter D. Morley

Abstract

This paper is concerned with the mathematical description of orbits that do not have a constant central gravitating mass. Instead, the attracting mass is a diffuse condensate, a situation which classical orbital dynamics has never encountered before. The famous Coma Cluster of Galaxies is embedded in Dark Matter. Condensed Neutrino Objects (CNO), which are stable assemblages of neutrinos and anti-neutrinos, are candidates for the Dark Matter. A CNO solution has been attained previously for the Coma Cluster, which allows mathematical modeling of galaxy orbital mechanics within Dark Matter, first reported here. For non-zero eccentricity galaxy orbits, each point along the trajectory sees a different gravitating central mass, akin to satellite orbits inside Earth. Mathematically, the galaxy orbits are non-Keplerian, spirographs.

Keywords: dark matter, coma cluster, condensed neutrino object, orbital dynamics, galaxy cluster

1. Introduction

There are two seminal observations that bracket the existence of Dark Matter, giving essential physics clues. The first is Zwicky [1] who noticed that the luminous matter in the Coma Galaxy Cluster is too small in mass to gravitationally bind the cluster. Quanitatively, the fastest bound galaxy has speed relative to the Coma center-of-mass of about 3000 km/s [2]. From the Coma Galaxy Cluster we learn:

1. The amount of (unseen) Dark Matter in the Coma Cluster vastly ‘outweighs’ the luminous matter.

2. We can see right through the Coma Cluster to image galaxies on the other side of the Universe, which means light is not scattered by Dark Matter: astonishingly, Dark Matter is transparent to light.

3. The gravitational potential of Dark Matter is of the same size as the Coma Galaxy Cluster dimensions.

The second seminal observation is Rubin [3] who showed that the rotational speeds of stars in spiral galaxies are too high for gravitational binding with the amount of luminous spiral galaxy mass observed. Unfortunately, here the story takes a tragic diversion, because Rubin assumed that the missing spiral Dark Matter

IntechOpen

must be in the halo of the measured galaxy itself. This mistake could be attributable to the lack of mathematical sophistication, but it has misled researchers for years. Let us discuss the situation of a spiral galaxy embedded in a Coma-like Galaxy Cluster Dark Matter potential and see the complexity of the resulting gravitational potential. The Dark Matter gravitational potential at position r inside the Dark Object is (G is the gravitational constant)

$$\Phi(\mathbf{r}) = -G\int\frac{\rho_B(\mathbf{r}')}{|\mathbf{r}-\mathbf{r}'|}d^3r' \tag{1}$$

where ρ_B is the Dark Matter mass density. For Coma-like Dark Matter, this density is spherically symmetric: $\rho_B(\mathbf{r}') = \rho_B(r')$ so Eq. (1) becomes

$$\Phi(\mathbf{r}_c + \mathbf{a}) = -G\int_0^{|\mathbf{r}_c+\mathbf{a}|}\frac{dM_B(r')}{r'} \tag{2}$$

where $dM_B(r') = \rho_B(r')4\pi r'^2 dr'$, $\mathbf{r}_c$ is the radius from the origin of the Dark Matter Object to the spiral galaxy center of mass and $\mathbf{a}$ is the spiral arm vector. The embedded spiral galaxy is shown in **Figure 1**. We're interested in the difference of gravitational potential between a spiral arm and its galaxy's center of mass, $\Phi(\mathbf{a}) = \Phi(\mathbf{r}_c + \mathbf{a}) - \Phi(\mathbf{r}_c)$.

$$\Phi(\mathbf{a}) = -G\int_{|\mathbf{r}_c|}^{|\mathbf{r}_c+\mathbf{a}|}\frac{dM_B(r')}{r'} \tag{3}$$

Consider now letting a move along a spiral arm, going around 360 degrees, where this angle becomes the spiral galaxy's azimuthal angle ϕ. If the spin axis is tilted with respect to a radial, then Eq. (3) has both positive and negative values: for some ϕ: $|\mathbf{r}_c + \mathbf{a}| > |\mathbf{r}_c|$ and for 180 degrees further in ϕ: $|\mathbf{r}_c + \mathbf{a}| < |\mathbf{r}_c|$. Circumlocution means a rotating star with a fixed distance from the spiral galaxy's center will go up

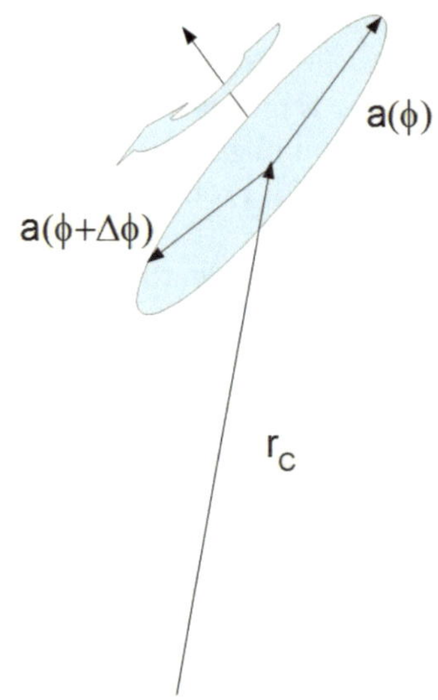

Figure 1.
Embedded spiral galaxy in a coma-like dark matter object with r_c the vector from object center to spiral center and a the spiral arm's position vector. The tilt of the spin vector s with respect to the dark matter radius vector $\mathbf{r}_c$ is the origin of the complex spiral arm star rotation speeds.

a potential hill for half its revolution and lose rotational speed. The other half of the rotation it will go down a potential hill and gain speed; the spiral arm rotational speeds become azimuthal angle dependent, Eq. (3). If however, the spin axis is parallel or antiparallel to a Dark Matter Object radial, then $|\mathbf{r}_c + \mathbf{a}|$ = constant $> |\mathbf{r}_c|$, and rotation speeds (for constant distance from center) are no longer azimuthally dependent from the Dark Matter gravitational potential. In the interesting other case that $\mathbf{a} \to 0$, very short distant scales (~ 1 Kpc), then there is no change in the Dark Matter potential ($\Phi(\mathbf{a}) \to 0$) and the star speeds are the *same near each other*. This geometric dependence of the large Coma-like Dark Matter is a significant source of confusion when the Dark Matter is attributable to a galactic halo and has led to bogus science claims [4] of satellite galaxies having enormous amounts of Dark Matter.

2. Dark matter particles as fermions

Large astronomical assemblages of matter have gravitational self-energies that will cause them to collapse. They are only stable if there is an internal pressure source. Observations of Dark Matter embedding galactic clusters reveal no Dark Matter energy generation. The Lamda-Cold-Dark-Matter cosmological model postulates that Dark Matter is cold. Therefore Maxwellian statistics are not present. Dark Matter particles are described by quantum statistics, which are either the Fermi-Dirac or Bose-Einstein distributions. However, boson stars do not exist in Nature because bosons will occupy the lowest energy state in cold matter. We come to the conclusion that Dark Matter is made up of Fermions that are in equilibrium due to degeneracy pressure. Recognizing that White Dwarfs are stable due to electron degeneracy, Neutron Stars are stable due to neutron degeneracy, we see that the size of the stable assemblage is inversely proportional to the Fermion mass. Zwicky's discovery that the Coma Cluster of galaxies is embedded in Dark Matter means that the Dark Matter Fermion particle must be incredibly small in mass, even compared to the small electron's mass.

3. Condensation of cosmological neutrinos

The additional requirement that the Dark Matter particles be the most abundent particles in the Universe identifies the condensation of cosmological neutrinos from the Big Bang as a very attractive candidate for Dark Matter. Reference [5] is the first publication that correctly evaluated the equation of state for degenerate neutrino matter, where the neutrinos and anti-neutrinos condense into stable assemblages called 'condensed neutrino objects' (CNO). Reference [6] derived the CNO mass-radius relationship

$$M(R) \simeq \frac{1.97462 \times 10^{15} M_\odot}{R^3 m_\nu^8} \tag{4}$$

where $M(R)$ is the mass of the stable CNO having radius R in units of Mpc and m_ν, neutrino mass scale (neutrinos are almost a perfect mass symmetry from neutrino mixing), is in units eV/c^2. Once m_ν is determined by the KATRIN terrestial experiment [7], Eq. (4) completely describes Dark Matter. Here, we take the opportunity to model galaxy orbital dynamics in the Coma Galaxy Cluster, using an estimated value for m_ν from reference [8].

The interesting physics of CNO is that there is no central mass. Instead, galaxies having non-zero eccentricities see a different gravitating mass at each point along their orbit. Human beings have never seen this astronomical phenomena before. It would be akin to having satellite orbits inside the Earth and it leads to non-Keplerian orbits.

4. Coma galaxy cluster CNO

When galaxies self-assemble inside CNO, they may have negligible velocities with respect to the Dark Matter or non-negligible velocities. Those galaxies having negligible velocities fall to the center of the CNO and execute simple harmonic motion (SHM), reference [5]. These galaxies then obtain their fastest speed at the center of the CNO, and when there, are the fastest galaxies embedded in the CNO that are gravitationally bound. On the other hand, galaxies which self-assemble with non-negligible velocities with respect to the CNO center of mass, execute orbital dynamics. Conservation of angular momentum prevents their appearance in the CNO center, and they never appear in the fast velocity histograms. If astronomical data is available for individual galaxies of a galaxy cluster, then picking out the fastest bound galaxy will place its location at or near the CNO center. In Ref. [8] this analysis was done to identify the CNO parameter, the neutrino Fermi Momentum (p_F) at the CNO center, associated with the Coma Cluster. The Fermi Momentum enters in the equation of hydrostatic equilibrium as $x = p_F/m_\nu c$ (called the reduced Fermi Momentum), where c is the speed of light. The different CNO in Eq. (4) have different boundary condition $x|_{R=0}$, which we denote by $x(0)$, after the neutrino mass scale is determined.

The Coma Galaxy Cluster CNO has solution [8] $x(0) = 0.010$ [1]. This Coma Cluster solution has mass $\mathcal{M}_{010}$ and radius $\mathcal{R}_{010}$

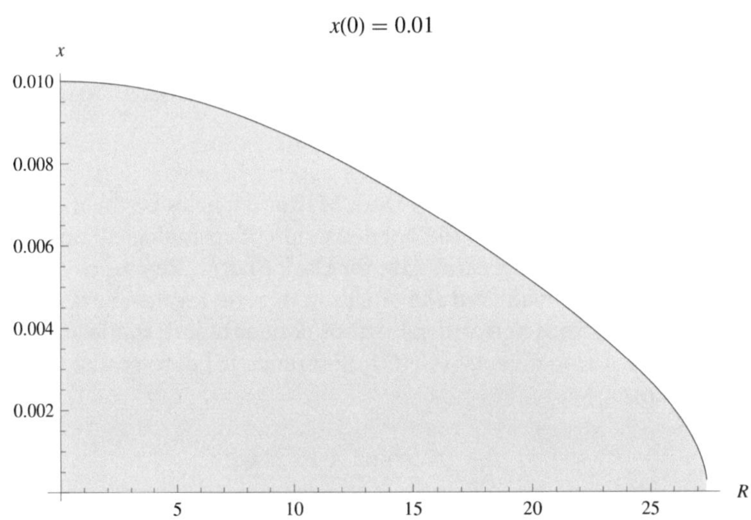

Figure 2.
Spatial variation of the reduced Fermi momentum x for the CNO having boundary value x(0) = 0.010. The units of length are 46128.98 pc/ m_ν^2 with m_ν in units of eV/c^2. The figure is from reference [8] and is the result of solving the equation of static equilibrium for degenerate matter.

[1] This shows just how non-relativistic these stable CNO are.

$$\mathcal{M}_{010} = \frac{9.809 \times 10^{14}}{m_\nu^2} M_\odot \tag{5}$$

$$\mathcal{R}_{010} = \frac{1.2625}{m_\nu^2} Mpc \tag{6}$$

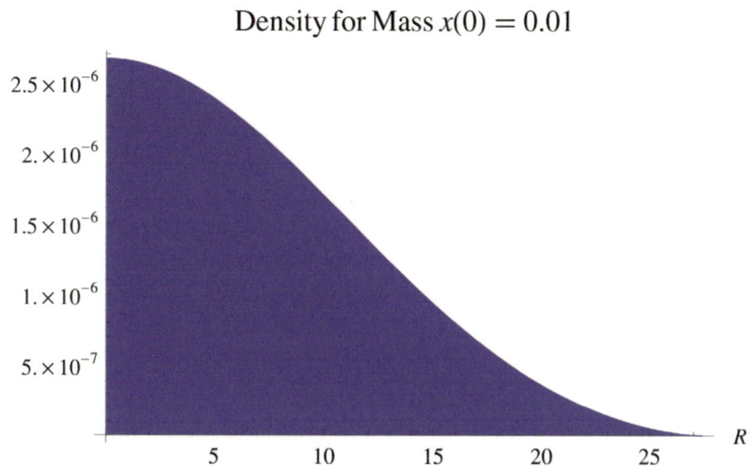

Figure 3.
CNO mass density for $x(0) = 0.010$ *CNO. The units of mass density are* $1.76307 \times 10^{-20} m_\nu^4$ *gm/cm*^{3} *with* m_ν *in units of* eV/c^2 *the figure is from reference [8]. Notice that the density is finite at the origin, which means there is no singularity present.*

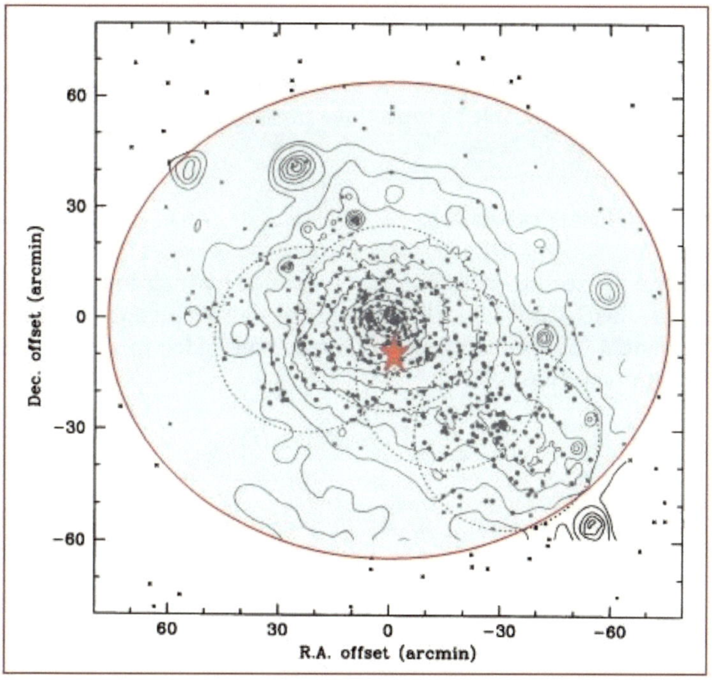

Figure 4.
Approximate location of the CNO center using fastest galaxy GMP = 3176, which has offsets R.a. +0.0579 arc-minutes and Dec −13.465 arc-minutes. The figure background is taken from reference [2], which shows the actual coma galaxy cluster and X-ray contour lines. At a distance of 101.3 Mpc, the 2.191 Mpc CNO radius translates to 74.35 arc-minutes. The composite figure is from reference [8].

where m_ν is in units of eV/c^2. In **Figure 2**, The Coma Galaxy Cluster CNO reduced Fermi momentum is plotted as a function of radial coordinate, while in **Figure 3**, the mass density is plotted.

In Ref. [8], $m_\nu = 0.759\ \mathrm{eV/c^2}$ was used. The center of the CNO embedding the Coma Galaxy Cluster was determined to be close to galaxy GMP [2] = 3176. The CNO Coma Galaxy Cluster solution is shown in **Figure 4**.

5. Galaxy cluster embedded in a CNO

Observationally, most or all galaxy clusters are embedded in Dark Matter, because the observable luminous matter does not produce a strong enough gravitational well to confine the experimentally observed galaxy velocities. This was the original Zwicky observation. Theoretical studies of galaxy cluster dynamical evolutions that did not understand the physics of Dark Matter reached the following conclusion (exemplified by reference [9]): galaxies collapse toward the center and virialize with Dark Matter to attain a steady-state distribution. For CNO Dark Matter, the galaxies revolve in a frictionless condensate and do not viralize with the Dark Matter at all.

5.1 Coma galaxy cluster dynamics

For the relevant neutrino mass scale of $m_\nu = 0.759\ \mathrm{eV/c^2}$, the Coma Dark Matter has mass 1.7 $\times 10^{15}$ M$_\odot$ with radius 2.191 Mpc. We will use this CNO to demonstrate how to mathematically compute orbits in CNO, using examples. These will be galaxies that have self-assembled with non-negligible velocities with respect to the CNO center-of-mass. The computation of the Dark Matter mass profile will allow for detailed simulations of galaxy cluster dynamics: formation of galaxies and their subsequent orbital evolution. In the section Future Work, it will be described how a simulation should be able to reproduce present measured kinetic velocity distributions.

5.2 Coma galaxy cluster orbits

The galaxies will execute orbits on a plane defined by their initial (birth) velocity components. On this 2-Dimensional plane are the r and θ polar coordinates. The radial and tangential force equations for a galaxy embedded in a CNO are (G is Newton's gravitational constant)

$$\ddot{r} - r\dot{\theta}^2 = \frac{-GM_{CNO}(r)}{r^2}$$

$$r\ddot{\theta} + 2\dot{r}\dot{\theta} = \frac{1}{r}\frac{d}{dt}\left(r^2\dot{\theta}\right) = 0 \qquad (7)$$

where $M_{CNO}(r)$ is the Dark Matter mass enclosed within radial cordinate r. For the Coma Galaxy Cluster solution of $x(0) = 0.010$ with $m_\nu = 0.759\ \mathrm{eV/c^2}$, this has the expression

$$M_{CNO}(r) = 5.5773407 \times 10^{17} M_\odot I(s) \qquad (8)$$

where $s = r/r_{CNO},\ \ s \leq 1$, with r_{CNO} = 2.1915 Mpc and $I(s)$ has a polynomial expansion

$$
\begin{aligned}
I(s) = &-0.000019832787608368 19 \cdot s + 0.00057121308662060 88 \cdot s^2 \\
&+0.012326181706900396 \cdot s^3 + 0.030069987017380402 \cdot s^4 \\
&-0.12115460629704124 \cdot s^5 + 0.13430437610312457 \cdot s^6 \\
&-0.06587537476021942 \cdot s^7 + 0.01283085557464332 \cdot s^8
\end{aligned} \tag{9}
$$

The numerical function $I(s)$ is displayed in **Figure 5**. In **Figure 6**, we show the orbital velocity components of a galaxy embedded in a CNO.

As already discussed, the reason why orbital dynamics of galaxies embedded in a CNO are different and unusual is because there is no central mass situated at the center of a CNO (see **Figure 5**). This is the first time we see non-Keplerian orbits from classical gravity. We now solve the system of Eq. (7) for interesting initial

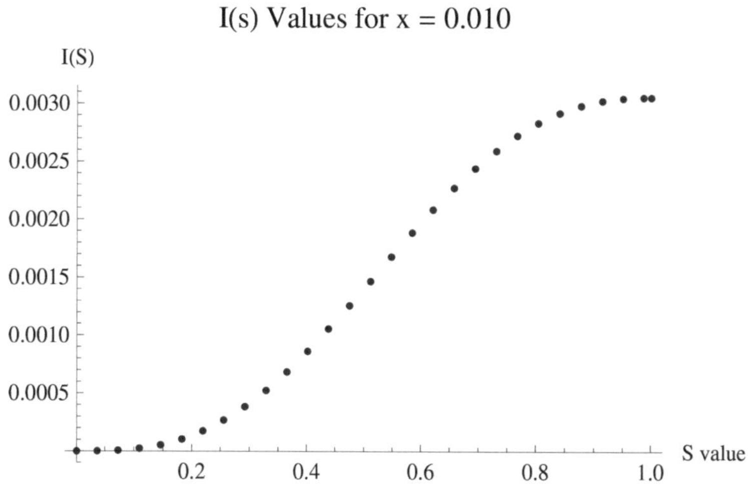

Figure 5.
Numerical evaluation of the function $I(s)$.

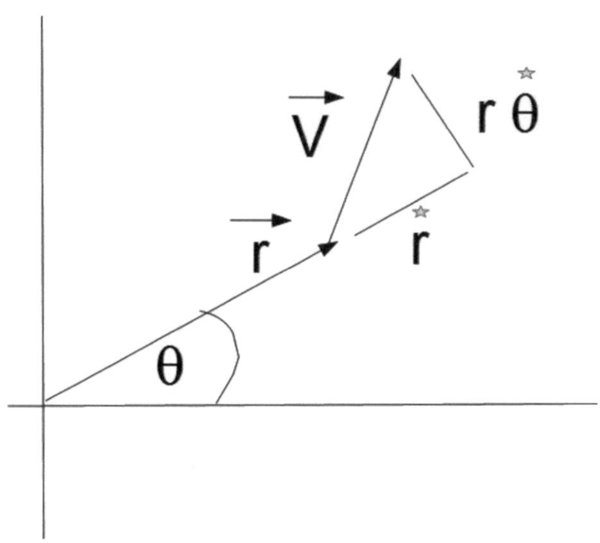

Figure 6.
2-dimensional (planar) velocity components for orbit dynamics of a galaxy embedded in dark matter CNO. The reference system is the center-of-mass of the CNO. The axis can be any perpendicular coordinates.

conditions. We will do two examples that illustrate the orbital dynamics of galaxies embedded in Dark Matter CNO.

The last equation of array Eq. (7) shows that the angular momentum per unit mass $h = r^2\dot{\theta}$ is a conserved quantity (a constant). Changing variables $r \rightarrow \tilde{s} = \frac{r_M}{r}$ where, for simplicity of notation, we put $r_{CNO} \equiv r_M$, one can show that the system Eq. (7) becomes

$$\frac{d^2\tilde{s}}{d\theta^2} + \tilde{s} = \frac{r_M G M_{CNO}(1/\tilde{s})}{h^2} \tag{10}$$

5.3 Case 1 example

The Coma Galaxy Cluster has a measured (in Coma center-of-mass frame) Gaussian distribution of speeds [2] that extends up to 3000 km/s. For our first example, we will use a small value for the initial speed of a galaxy: V_0 = 100 km/s with a velocity angle 75 degrees, starting at $\theta = 0$ degrees and located at $r_{initial} = .2r_M$. This gives the initial velocity condition (see **Figure 6**)

$$\dot{r}|_{initial} = V_0 \cos 75^\circ \quad = \quad 25.8819\ldots \text{km/s} \tag{11}$$

$$(r\dot{\theta})\big|_{initial} = V_0 \sin 75^\circ \quad = \quad 96.5925\ldots \text{km/s} \tag{12}$$

For $r_{initial} = .2r_M \rightarrow \tilde{s}|_{initial} = 5$, so we have the first initial condition. We next use the chain rule of calculus

$$\dot{r} = \frac{dr}{d\theta}\dot{\theta} = \frac{h}{r^2}\frac{dr}{d\theta} \tag{13}$$

Since $\tilde{s} = r_M/r$, we have using Eq. (13)

$$\frac{d\tilde{s}}{d\theta} = -\frac{r_M}{r^2}\frac{dr}{d\theta} = -\frac{r_M}{h}\dot{r} \tag{14}$$

This gives the last initial condition

$$\left.\frac{d\tilde{s}}{d\theta}\right|_{initial} = -\frac{r_M}{h}\dot{r}_{initial} \tag{15}$$

Evaluating Eq. (15) gives

$$\left.\frac{d\tilde{s}}{d\theta}\right|_{initial} = -1.339745962 \tag{16}$$

Designating the right hand side of Eq. (10) as α, we get

$$\alpha = \frac{r_M G M_{CNO}}{h^2} = 2932920.356 \cdot I(1/\tilde{s}) \tag{17}$$

Finally, the problem to be numerically solved is reduced to

$$\frac{d^2\tilde{s}}{d\theta^2} = -\tilde{s} + 2932920.356 \cdot I(1/\tilde{s}) \tag{18}$$

$$\tilde{s}|_{initial} = 5 \tag{19}$$

$$\left.\frac{d\tilde{s}}{d\theta}\right|_{initial} = -1.339745962 \tag{20}$$

In **Figure 7** we give the polar graph of 4 revolutions showing that the orbit precesses without any relativistic corrections or perturbations. This is non-Keplerian behavior: a spiralgraph. If we look closely at the starting location $r|_{initial} = .2r_M$ with $\theta = 0$, we see that the galaxy does NOT extend beyond its inital radius, with the next step in polar angle showing that it is closer to the CNO center. The reason for this behavior is that the starting speed of 100 km/s at $r|_{inital} = .2r_M$ is too small for the galaxy to extend its orbit for the next discrete angle value. When we do the second example, where we change the starting speed from 100 km/s $\rightarrow$ 1000 km/s, we will see that the galaxy expands beyond $r|_{initial}$. Another aspect of **Figure 7**: it alludes to the fact that galaxies which assemble with negligible velocities undergo simple harmonic motion [5].

We now compute the period for this case. In order to do this, we have to arrange the variables such that

$$\frac{d\theta}{dt} = f(\theta) \tag{21}$$

so the period τ is

$$\tau = \int_0^{2\pi} \frac{d\theta}{f(\theta)} \tag{22}$$

This is done using h = constant = $r^2 \frac{d\theta}{dt}$. Working this out

$$\frac{d\theta}{dt} = f(\theta) = \tilde{s}^2 \left(2.856813144 \times 10^{-19}\right)\ \text{s}^{-1} \tag{23}$$

This gives

$$\tau\ \text{Case1} = \int_0^{2\pi} \frac{d\theta}{\tilde{s}^2(\theta)} (110.9970809)\ \text{Gyr} \tag{24}$$

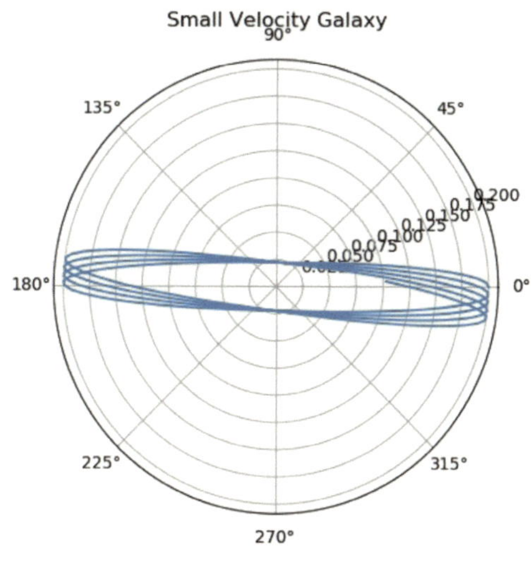

Figure 7.
Orbital mechanics for small velocity coma cluster galaxy.

Doing this final integration numerically reveals

$$\tau \text{ Case1} = 3.2502 \text{ Gyr} \tag{25}$$

In **Figure 8**, the speed is plotted against rotation angle, showing the near SHM of small birth velocities. Starting at 100 km/s at 0 degrees, it reaches a high ~ 870 km/s at closest approach to the center and then back again to 100 km/s on the other side, for one-half revolution.

5.4 Case 2 example

The only change from the prior case is the initial speed V_0: 100 km/s $\rightarrow$ 1000 km/s. One can show that the problem to solve becomes

$$\frac{d^2\tilde{s}}{d\theta^2} = -\tilde{s} + 29329.20356 \cdot I(1/\tilde{s}) \tag{26}$$

$$\tilde{s}|_{initial} = 5 \tag{27}$$

$$\left.\frac{d\tilde{s}}{d\theta}\right|_{initial} = -1.339745962 \tag{28}$$

In **Figure 9**, we give 4 revolutions for this case, again showing non-Keplerian behavior. Note that the starting initial conditions permit the galaxy to expand beyond $r|_{initial} = .2r_M$.

For the period of this case, one can show that the numerical integration is

$$\tau \text{ Case2} = \int_0^{2\pi} \frac{d\theta}{\tilde{s}^2(\theta)}(11.09970809) \text{ Gyr} \tag{29}$$

Doing this final integration numerically finds

$$\tau \text{ Case2} = 3.1765 \text{ Gyr} \tag{30}$$

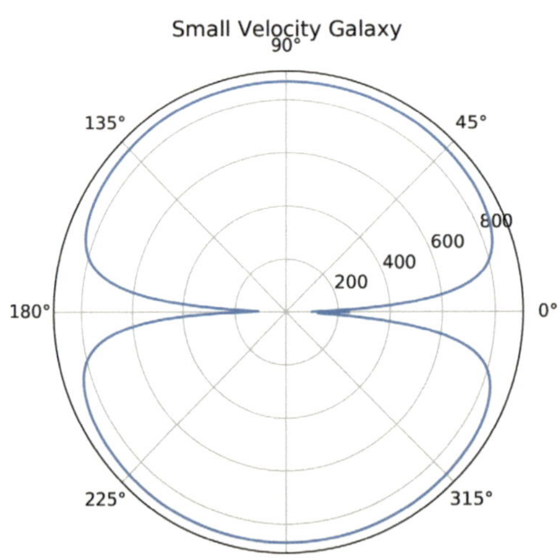

Figure 8.
The speed of the small velocity galaxy for one revolution, showing the near SHM for small initial birth velocities. See text for explanation.

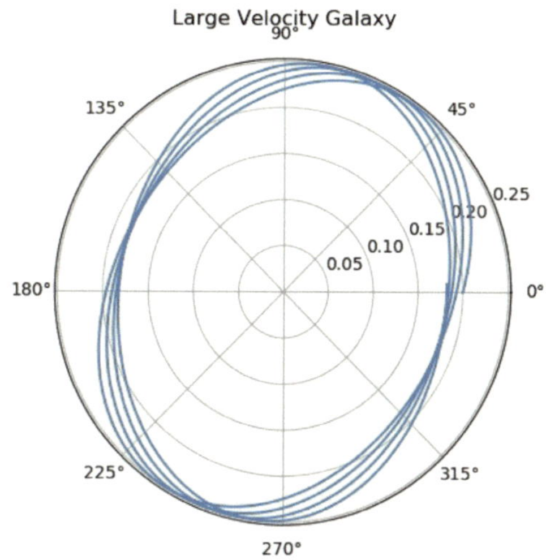

Figure 9.
Orbital mechanics for large velocity coma cluster galaxy.

6. Conclusion

The KATRIN neutrino mass experiment [7] can prove Dark Matter is the condensation of cosmological neutrinos and anti-neutrinos by obtaining a mass signal for the electron-anti-neutrino. The Planck Satellite Consortium assumes no condensation of cosmological neutrinos in their analysis of the cosmological microwave background and predicts neutrino masses too small for KATRIN to measure [8]. The identification of CNO as the Dark Matter allows mathematical modeling of embedded galaxy orbits, first reported here, using the Coma Cluster of galaxies.

7. Future work

Galaxies inside the CNO Dark Matter self-aggregate with a probability distribution at locations between $\vec{r}$ and $\vec{r} + \vec{d}r$ and with initial velocities between $\vec{v}$ and $\vec{v} + \vec{d}v$. For example, if the probability of assemblage between r and $r + dr$ is the fractional volume of available space in the CNO, then the average initial radius distance would be $.75 \cdot r_{CNO}$. The cluster galaxies then perform orbits, or collapsing toward the center. There is no equilibrium with respect to the Dark Matter, as the baryons move through a frictionless condensate. The periods have units Gyr, so a present snapshot is good for a long time.

The next step is to do a simulation of N-number of self-aggregating galaxies over an initial period of time and time-advance it to the present day. The 'birth' distribution in velocities will give rise to a predicted later time-evolved velocity distribution that can be compared to the present-day astronomically measured velocity distribution.

Because of the huge CNO Dark Matter mass, individual galaxy collisions are a small perturbation of cluster dynamics. The probability that you have cluster evaporation from many-body interactions is near zero and completely negligible: baryonic matter situated in a CNO gravitational well stays in the CNO gravitational well. For our numerical examples here, we used two galaxies at the same initial radial distance. However, the orbital planes of both galaxies could have been in any

3-dimensional orientation in 3-space, hence even the same initial radial orbits have a negligible chance of interacting.

Conflict of interest

The author states that there is no conflict of interest.

Abbreviations and notation

CNO	condensed neutrino object, stable assemblage of neutrinos and anti-neutrinos.
G	Newton's gravitational constant.
$\Phi(\mathbf{r})$	Dark Matter gravitational potential (potential energy per unit mass) at location r, Eq. (1).
$\mathbf{r}_c$	Spiral galaxy center-of-mass position inside the Dark Matter Object.
$\mathbf{a}$	spiral arm vector in spiral galaxy center of mass.
$M(R)$	mass of CNO, Eq. (4)
m_ν	neutrino mass scale, to be measured by [7].
R	radius of CNO, Eq. (4)
SHM	simple harmonic motion.
$x = p_F/m_\nu c$	reduced Fermi momentum, where c is the speed of light.
$\mathcal{M}_{010}$, $\mathcal{R}_{010}$	modeled mass and radius of the Coma Galaxy cluster CNO, Eqs. (5) and (6).
Mpc	mega-par-sec.
$M_\odot$	mass of the Sun.
$M_{CNO}(r)$	enclosed CNO mass at radius r, Eq. (8).
$\tilde{s} = r_M/r$,	where $r_M \equiv r_{CNO}$ is the radius of the Coma CNO.

Author details

Peter D. Morley
Blue Ridge Scientific LLC, Front Royal, VA, United States

*Address all correspondence to: peter3@uchicago.edu

References

[1] Zwicky F. On the masses of nebulae and of clusters of nebulae. Ap. J. 1937; **86**:217-246

[2] Colless M, Dunn AM. Structure and dynamics of the coma cluster. Ap. J. 1996;**458**(435). DOI: 10.1086/176827

[3] Vera C. Rubin and W. Kent Ford, Jr. Rotation of the Andromeda Nebula from a spectroscopic survey of emission regions. Ap. J. 159, 379-403 (1970).

[4] Chiti A et al. An extended halo around an ancient Dwarf galaxy. Nat. Astron. 2021. DOI: 10.1038/s41550-020-01285-w

[5] Morley PD, Buettner DJ. SHM of galaxies embedded within condensed neutrino matter. Intl. Journ. Mod. Phys. 2015;**24**:1550004. DOI: 10.1142/S0218271815500042

[6] Morley PD, Buettner DJ. A dark matter signature for condensed neutrinos. Intl. Journ. Mod. Phys. 2016; **25**:1650089. DOI: 10.1142/s0218271816500899

[7] KATRIN. https://www.katrin.kit.edu (2019).

[8] Peter D. Morley. Prediction of the neutrino mass scale using Coma Galaxy Cluster data. Symmetry. 2020;**12**(1049). DOI: 10.3390/sym12061049

[9] Andrey V. Kravtsov and Stefano Borgani. Formation of galaxy clusters. Ann. Rev. Astro. Astrophy. 2012;**50**: 353-409

Chapter 7

Cosmology and Cosmic Rays Propagation in the Relativity with a Preferred Frame

Georgy I. Burde

Abstract

In this chapter, cosmological models and the processes accompanying the propagation of the cosmic rays on cosmological scales are considered based on particle dynamics, electrodynamics and general relativity (GR) developed from the basic concepts of the 'relativity with a preferred frame'. The 'relativity with a preferred frame', designed to reconcile the relativity principle with the existence of the cosmological preferred frame, incorporates the preferred frame at the fundamental level of special relativity (SR) while retaining the fundamental space-time symmetry which, in the standard SR, manifests itself as Lorentz invariance. The cosmological models based on the modified GR of the 'relativity with a preferred frame' allow us to explain the SNIa observational data without introducing the dark energy and also fit other observational data, in particular, the BAO data. Applying the theory to the photo pion-production and pair-production processes, accompanying the propagation of the Ultra-High Energy Cosmic Rays (UHECR) and gamma rays through the universal diffuse background radiation, shows that the modified particle dynamics, electrodynamics and GR lead to measurable signatures in the observed cosmic rays spectra which can provide an interpretation of some puzzling features found in the observational data. Other possible observational consequences of the theory, such as the birefringence of light propagating in vacuo and dispersion, are discussed.

Keywords: general relativity, FRW models, late-time cosmic acceleration, dark energy, UHECR, gamma rays, photo pion-production, pair-production

1. Introduction

Lorentz symmetry is arguably the most fundamental symmetry of physics, at least in its modern conception. Physical laws are Lorentz-covariant among inertial frames; namely, the form of a physical law is invariant under the Lorentz group of space-time transformations. Therefore, the Lorentz symmetry sets a fundamental constraint for physical theories. Nevertheless, modifications of special relativity (SR) and possible violations of Lorentz invariance have recently obtained increased attention. Although, the success of general relativity (GR) to describe all observed gravitational phenomena proves the fundamental importance of Lorentz invariance in our current understanding of gravitation, some of the modern theories (unification theories, extensions of the standard model and so on) suggest a violation of special relativity. The aim of most of the Lorentz violating theories is to modify a Lorentz invariant theory by introducing small phenomenological Lorentz-violating terms into the basic relations of the theory (Lagrangian density, dispersion relation and so on) and predict what can be expected from it. Reviews of the most popular approaches [1–26] to parameterizing Lorentz violating physics in the context of their relation to the 'relativity with a preferred frame' can be found in [27, 28]. Some of those studies are discussed in the following sections about the results obtained in the present paper.

The theory termed 'relativity with a preferred frame' developed in [27–29] represents a very special type of a Lorentz violating theory that is conceptually different from others found in the literature. It is not even a preferred frame that makes a difference—all violations of Lorentz invariance, made by distorting Lorentz-invariant relations of the theory, imply the existence of a preferred frame for the formulation of the physical laws, the one in which all the calculations need to be carried out, since breaking relativistic invariance also invalidates the transformations that allow us to change reference frame. The first major difference of the present analysis from the above-mentioned studies is that the Lorentz violation *is not introduced* into the theory but it is a result of using freedom in formulation one of two basic principles of special relativity, the principle of universality of the speed of light. In other terms, Lorentz's violation is ingrained into the framework of the theory at some fundamental level. The second major difference is that the relativistic invariance, in the sense that the form of a physical law is invariant under the space-time transformations between inertial frames, is not violated—it is a *Lorentz violation without violation of relativistic invariance*.

To outline the framework of the theory named 'relativity with a preferred frame' one has to start from the definition of the preferred frame. In the 'relativity with a preferred frame', the preferred frame is defined as the only frame where propagation of light is isotropic, while it is anisotropic in all other frames moving relative to the preferred one (it is a common definition in the studies investigating the fundamentals of special relativity and its potential breaking).[‡] Discussing the anisotropy of propagation of light one has to distinguish between the *two-way* speed of light, i.e. the average speed from source to observer and back, and the *one-way* speed which is a speed of light in one direction—*either* from source to observer *or* back. In the 'relativity with a preferred frame', it is the *one-way* speed of light that is assumed to be anisotropic in all the frames except the preferred frame, while the *two-way* speed of light is isotropic and equal to c in all inertial frames.[§] The analysis is based on the invariance of the equation of (anisotropic) light propagation for the space-time transformations between inertial frames and the group structure of the transformations plays a central role in the analysis. Although, the existence of the preferred frame seems to be in contradiction both with the basic principles of special relativity and with the group property of the transformations, in the framework of the 'relativity with a preferred frame', those principles are retained. The crucial element, which allows retaining the relativistic invariance and the group

[‡] It is worth noting that, although the anisotropy of speed of light is one of the central features of the present analysis, this theory stands apart from the ample literature on the conventionality of simultaneity and clock synchronization. A discussion of those issues in the context of the 'relativity with a preferred frame' can be found in [29, 30].

[§] In the modern versions of the experiments designed to test special relativity and the so-named 'test theories' (e.g., [31, 32], see a discussion in [27, 29, 30]), the tests are meant to detect the anisotropy of the *two-way* speed of light.

property of the space-time transformations, is that the anisotropy parameter k, figuring in the equation of the anisotropic light propagation, is treated as a variable that takes part in the group transformations (for more details, see Section 2). Then the preferred frame, in which $k = 0$, enters the analysis on equal footing with other frames since nothing distinguishes the transformations to/from that frame from the transformations between two frames with $k \neq 0$. The space-time symmetry underlying the group of transformations between inertial frames, which in the standard SR is expressed by the existence of the combination invariant under the transformations (interval), in the 'relativity with a preferred frame', reveals itself also in the form of the invariant combination, a counterpart of the interval of the standard SR. Such a 'modified space-time symmetry' paves the way to extensions of the kinematics of the 'relativity with a preferred frame' to free-particle dynamics, general relativity and electromagnetic field theory.

The above-described generalization of special relativity cannot be validated by experiments measuring the speed of light since only the two-way speed of light, the same in all the frames, can be measured. For creating a physical theory, predictions of which can be compared with observational data, it is needed to identify the preferred frame of the present analysis, which is defined by the property of isotropy of the one-way speed of light, with a frame possessing the property that velocity of any other frame relative to it can be measured using some physical phenomena. In the present analysis, that preferred frame is a comoving frame of cosmology or the CMB frame (note that identifying the preferred frame with the CMB frame is a common feature of practically all Lorentz-violating theories). It is the only frame possessing the property, that motion of any other frame relative to it is distinguishable, and, in addition, this frame, like the preferred frame of the present analysis, is defined based on the isotropy property. As a result of specifying the preferred frame, all the relations of the 'relativity with a preferred frame', as well as of its extensions, contain only one universal constant b which is a parameter to be adjusted for fitting the results of the theory to observational data.

Identifying the preferred frame with the cosmological comoving frame implies that the theory should be applied to phenomena on cosmological scales. Studying different phenomena requires extensions of the modified SR kinematics to different areas of physics. The purpose of this chapter is to present a unified view of the extensions and their applications based on the concept of the modified space-time symmetry. This includes extension to general relativity (Section 4.1) and constructing cosmological models based on the modified general relativity (Section 4.2); extension to the dynamics of the free particles (Section 3.1) and its application to the processes accompanying the Ultra High Energy Cosmic Rays (UHECR) and the gamma-rays propagation (Sections 5.1 and 5.2); extension to electromagnetic field (Section 3.2) and studying electromagnetic waves based on the modified electrodynamics (Section 3.3) with application to the gamma-rays propagation (Section 5.3).

2. Special relativity kinematics

Kinematics of the 'relativity with a preferred frame' will be only outlined in this section, for a detailed presentation see [27–29].

The transformations between two arbitrary inertial reference frames S and S', with the coordinate systems $\{X, Y, Z, T\}$ and $\{x, y, z, t\}$ in the standard configuration (with the y- and z-axes parallel to the Y- and Z-axes and S' moving relative to S with the velocity v in the positive direction of the common x-axis), are considered. In the subsequent analysis, the group property of the space-time transformations is used as a primary tool. Groups of transformations are sought using the condition of invariance of the equation of anisotropic light propagation [30]

$$ds^2 = c^2dt^2 - 2kc\ dtdx - \left(1 - k^2\right)dx^2 - dy^2 - dz^2 = 0 \tag{1}$$

where k is the anisotropy parameter such that speeds of light in the positive and negative x -directions are

$$c^{(+)} = \frac{c}{1+k}, \quad c^{(-)} = \frac{c}{1-k} \tag{2}$$

Eq. (1) incorporates both the anisotropy of the *one-way* speed of light as equation (2) shows and the universality of the *two-way* speed of light in the sense that it is equal to c in all inertial frames (see, e.g., [33, 34]).[‖] The transformations involve both the space and time coordinates (x, y, z, t) and the anisotropy parameter k so that the equations of light propagation in the frames S and S' are

$$c^2dT^2 - 2Kc\ dTdX - \left(1 - K^2\right)dX^2 - dY^2 - dZ^2 = 0, \tag{3}$$

$$c^2dt^2 - 2kc\ dtdx - \left(1 - k^2\right)dx^2 - dy^2 - dz^2 = 0 \tag{4}$$

where K and k are the values of the anisotropy parameter in the frames S and S' respectively. The one-parameter (a) group of transformations of variables from $\{X, Y, Z, T, K\}$ to $\{x, y, z, t, k\}$, which converts (3) into (4), is sought in the form

$$\begin{aligned} &x = f(X, T, K; a), \quad t = q(X, T, K; a); \\ &y = g(Y, Z, K; a), \ z = h(Y, Z, K; a); \quad k = p(K; a) \end{aligned} \tag{5}$$

where, based on the symmetry arguments, it is assumed that the transformations of the variables x and t do not involve the variables y and z and vice versa. According to the Lie group method (see, e.g., [35, 36])., the infinitesimal transformations corresponding to (5) are introduced, as follows

$$\begin{aligned} &x \approx X + \xi(X, T, K)a, \quad t \approx T + \tau(X, T, K)a, \\ &y \approx Y + \eta(Y, Z, K)a, \quad z \approx Z + \zeta(Y, Z, K)a, \quad k \approx K + \kappa(K)a \end{aligned} \tag{6}$$

Proceeding by the usual Lie group technique (see [27–29] for details) one can define the form of the transformations in (x, y, z, t, k) variables. Calculating invariants of the group one can define a combination (a counterpart of the interval of the standard relativity) that is invariant under the transformations, namely

$$d\tilde{s}^2 = \frac{1}{\lambda(k)^2}\left(c^2dt^2 - 2kc\ dtdx - \left(1 - k^2\right)dx^2 - dy^2 - dz^2\right) \tag{7}$$

‖ Although the form (1) seems to be attributed to the one-dimensional formulation, in the three-dimensional case, the equation has the same form if the anisotropy vector **k** is directed along the x-axis [30]. In the present analysis, the x-axis defines also the line of relative motion of the two frames but it does not lead to any ambiguity. The assumption, that the anisotropy vector **k** is along the direction of relative motion of the frames S' and S, is justified by that one of the frames in a set of frames with different values of k is a preferred frame. Since the anisotropy is attributed to the motion with respect to the preferred frame, it is expected that the axis of anisotropy is either in the direction of motion or opposite to it.

where

$$\lambda(k) = exp\left[-\int_0^k \frac{p}{\kappa(p)}dp\right] \tag{8}$$

with $\kappa(k)$ being the group generator for the variable $k(a)$, see equation (6). Furthermore, introducing the new variables

$$\widetilde{t} = \frac{1}{c\lambda(k)}(ct - kx), \quad \widetilde{x} = \frac{1}{\lambda(k)}x, \quad \widetilde{y} = \frac{1}{\lambda(k)}y, \quad \widetilde{z} = \frac{1}{\lambda(k)}z \tag{9}$$

converts the invariant combination (7) into the Minkowski interval

$$d\widetilde{s}^2 = c^2 d\widetilde{t}^2 - d\widetilde{x}^2 - d\widetilde{y}^2 - d\widetilde{z}^2 \tag{10}$$

while the transformations take the form of rotations in the $(\widetilde{x},\widetilde{t})$ space (Lorentz transformations). However, in the calculation of physical effects, the 'true' time and space intervals in the 'physical' variables (t,x,y,z), obtained from $(\widetilde{t},\widetilde{x},\widetilde{y},\widetilde{z})$ by the transformation inverse to (9), are to be used.

The expression (7) for the modified interval and the transformations (9) contain the function $\lambda(k)$ which depends on the unspecified function $\kappa(k)$, the infinitesimal group generator for the variable k. This uncertainty reflects the fact that, within the above-developed framework, there is no possibility to determine the value of the anisotropy parameter k or, in other terms, to determine which frame is the preferred one, since only the two-way speed of light, equal to c in all the frames, can be measured. To specify the theory, such that its predictions could be compared with observations, there should exist a possibility to measure the frame velocity relative to a preferred frame using some other physical phenomena. Under the assumption that it is possible, the argument, that anisotropy of the one-way speed of light in an arbitrary inertial frame is due to its motion for a preferred frame, combined with group properties of the transformations, leads to the conclusion that the anisotropy parameter k in a frame moving relative to a preferred frame with velocity $\bar{\beta} = \bar{v}/c$ should be given by some universal function of that velocity, as follows

$$k = F(\bar{\beta}) \quad \text{or} \quad \bar{\beta} = f(k) \tag{11}$$

where $\bar{\beta} = f(k)$ is a function inverse to $F(\bar{\beta})$. Then the group generator $\kappa(k)$ is calculated by (see [27–29] for details)

$$\kappa(k) = \frac{1 - f^2(k)}{f'(k)} \tag{12}$$

which allows to calculate the factor $\lambda(k)$ from (8). Next, with the expression (11) for k introduced into (8), the factor $\lambda(k)$ becomes a function $B(\bar{\beta})$ of the frame velocity $\bar{\beta}$ relative to a preferred frame, as follows

$$\lambda(k(\bar{\beta}) \Rightarrow B(\bar{\beta}) = exp\left[-\int_0^{\bar{\beta}} \frac{F(m)}{1 - m^2}dm\right] \tag{13}$$

In the subsequent analysis, those general relations are specified using an approximation for $F(\bar{\beta})$ based on the following argument. An expansion of the function $F(\bar{\beta})$ in series in $\bar{\beta}$ should not contain even powers of $\bar{\beta}$ since it is expected that a direction of the anisotropy vector changes to the opposite if a direction of motion for a preferred frame is reversed: $F(\bar{\beta}) = -F(-\bar{\beta})$. Thus, with accuracy up to the third order in $\bar{\beta}$, the dependence of the anisotropy parameter on the velocity for a preferred frame can be approximated by

$$k = F(\bar{\beta}) \approx b\bar{\beta}, \quad \bar{\beta} = f(k) \approx k/b \tag{14}$$

With this approximation, the group generator $\kappa(k)$ calculated using (12) takes the form

$$\kappa(k) = b - \frac{k^2}{b} \tag{15}$$

and, correspondingly, the factors $\lambda(k)$ and $B(\bar{\beta})$ calculated from equations (8) and (13) become

$$\lambda(k) = \left(1 - \frac{k^2}{b^2}\right)^{b/2} \tag{16}$$

$$B(\bar{\beta}) = \left(1 - \bar{\beta}^2\right)^{b/2} \tag{17}$$

Thus, after the specification, all the equations contain only one undefined parameter, a universal constant b. It is worth reminding that, even though the specified law (14) is linear in β, it does include the second-order term which is identically zero. Therefore describing the anisotropy effects, which are of the order of β^2, by the above equations, is legitimate. In particular, the expression (17) for $B(\bar{\beta})$ is valid up to the second-order in β and, with the same order of approximation, it can be represented as

$$B(\bar{\beta}) = 1 - \frac{b}{2}\bar{\beta}^2 \tag{18}$$

3. Extensions to other areas of physics

3.1 Free particle dynamics

In this section, the free particle dynamics of the 'relativity with a preferred frame' developed in [28] is presented in a shortened form. The modified dynamics is developed based on the existence of the invariant combination $d\tilde{s}$ (a counterpart of the interval of the standard SR) defined by equation (7). Then the action integral for a free material particle is [37]

$$S = -mc\int_a^b d\tilde{s} = \int_{t_a}^{t_b} L dt \tag{19}$$

where the integral is along the world line between two given world points and L represents the Lagrange function. The invariant $d\tilde{s}$ defined by (7) can be represented in the form

$$d\tilde{s} = \frac{cdtQ(k, \beta_x, \beta)}{\lambda(k)} \tag{20}$$

where

$$Q(k,\beta_x.\beta) = \sqrt{(1-k\beta_x)^2 - \beta^2}; \qquad \beta_x = \frac{v_x}{c}, \quad \beta^2 = \frac{v_x^2 + v_y^2 + v_z^2}{c^2} \tag{21}$$

and

$$v_x = \frac{dx}{dt}, \; v_y = \frac{dy}{dt}, \; v_z = \frac{dz}{dt} \tag{22}$$

are components of the velocity vector. Then the Lagrange function is defined by

$$L = -mc^2 \frac{Q(k,\beta_x.\beta)}{\lambda(k)} \tag{23}$$

which is used to obtain expressions for the momentum $\boldsymbol{P}$ and energy E of a particle, as follows

$$P_x = \frac{1}{c}\frac{\partial L}{\partial \beta_x} = mc\frac{k + \beta_x(1-k^2)}{\lambda(k)Q(k,\beta_x.\beta)}, \quad P_y = \frac{1}{c}\frac{\partial L}{\partial \beta_y} = mc\frac{\beta_y}{\lambda(k)Q(k,\beta_x.\beta)},$$
$$P_z = \frac{1}{c}\frac{\partial L}{\partial \beta_z} = mc\frac{\beta_z}{\lambda(k)Q(k,\beta_x.\beta)} \tag{24}$$

and

$$E == P_x v_x + P_y v_y + P_z v_z - L = mc^2 \frac{1-k\beta_x}{\lambda(k)Q(k,\beta_x.\beta)} \tag{25}$$

Proceeding with the four-dimensional formulation, we will use the variables $(\tilde{t},\tilde{x},\tilde{y},\tilde{z})$ defined by (9) which allows converting the invariant combination (7) into the form (10) of the Minkowski interval. Introducing the four-dimensional contrainvariant radius vector by

$$\left(x^0,x^1,x^2,x^3\right) = (c\tilde{t},\tilde{x},\tilde{y},\tilde{z}) = \frac{1}{\lambda(k)}(ct-kx,x,y,z) \tag{26}$$

we define the contrainvariant four-velocity vector as

$$u^i = \frac{dx^i}{d\tilde{s}} \tag{27}$$

where the superscript i runs from 0 to 3. Using (26) and (20) in (27) yields

$$\left(u^0,u^1,u^2,u^3\right) = \frac{1}{Q(k,\beta_x.\beta)}\left(1-k\beta_x,\beta_x,\beta_y,\beta_z\right) \tag{28}$$

where $Q(k,\beta_x.\beta)$ is defined by (21). Correspondingly, covariant four-dimensional radius-vector and velocity vector are defined by

$$(x_0,x_1,x_2,x_3) = (c\tilde{t},-\tilde{x},-\tilde{y},-\tilde{z}), \tag{29}$$

$$(u_0,u_1,u_2,u_3) = \frac{1}{Q(k,\beta_x.\beta)}\left(1-k\beta_x,-\beta_x,-\beta_y,-\beta_z\right) \tag{30}$$

and the following relations hold

$$dx_i dx^i = d\tilde{s}^2 \tag{31}$$

$$u^i u_i = 1 \tag{32}$$

where a common rule of summation over repeated indexes is assumed. Next, recalling that the momentum four-vector is defined by

$$p_i = -\frac{\partial S}{\partial x^i} \tag{33}$$

and using the principle of the least action [37] we find (see [28] for details) that

$$p_i = mcu_i \tag{34}$$

while the contravariant components of the four-momentum vector are

$$p^i = mcu^i \tag{35}$$

Then from the identity (32) we get

$$p_i p^i = m^2 c^2 \tag{36}$$

Recalling that

$$P_x = \frac{\partial S}{\partial x}, \quad P_y = \frac{\partial S}{\partial y}, \quad P_z = \frac{\partial S}{\partial z}, \quad E = -\frac{\partial S}{\partial t} \tag{37}$$

with allowance for (26) and (33), we have

$$\begin{gathered} P_x = \frac{1}{\lambda(k)}\left(\frac{\partial S}{\partial x^1} - k\frac{\partial S}{\partial x^0}\right) = \frac{kp_0 - p_1}{\lambda(k)}, \quad P_y = \frac{1}{\lambda(k)}\frac{\partial S}{\partial x^2} = -\frac{p_2}{\lambda(k)}, \\ P_z = \frac{1}{\lambda(k)}\frac{\partial S}{\partial x^3} = -\frac{p_3}{\lambda(k)}, \quad E = -\frac{c}{\lambda(k)}\frac{\partial S}{\partial x^0} = \frac{cp_0}{\lambda(k)} \end{gathered} \tag{38}$$

which, upon using (34) and (30), yields the relations (24) and (25) for the three-momentum and energy. Solving equations (38) for the components of the four-momentum vector we get

$$p_0 = \frac{E\lambda(k)}{c}, \quad p_1 = \lambda(k)\left(\frac{kE}{c} - P_x\right), \quad p_2 = -\lambda(k)P_y, \quad p_3 = -\lambda(k)P_z \tag{39}$$

Then using (39) in (36) yields a dispersion relation for a free particle which can be represented in the form

$$\left(\frac{E}{c^{(+)}} - P_x\right)\left(\frac{E}{c^{(-)}} + P_x\right) = P_y^2 + P_z^2 + \frac{m^2c^2}{\lambda(k)^2} \tag{40}$$

where the speeds of light $c^{(+)}$ and $c^{(-)}$ in the positive and negative x -directions are defined by equation (2). It follows from (40) that for massless particles moving along the x -axis in the positive x direction

$$P_x = \frac{E}{c^{(+)}} = \frac{E(1+k)}{c} \tag{41}$$

while for massless particles moving in the negative x direction

$$P_x = -\frac{E}{c^{(-)}} = -\frac{E(1-k)}{c} \tag{42}$$

3.2 Electromagnetic field equations

The invariant action integral for a charged material particle in the electromagnetic field is made up of two parts: the action for the free particle defined by (19) and a term describing the interaction of the particle with the field. The invariance is provided by using the combinations that are invariant in the Minkowskian variables (26) so that the action integral takes the form [37]

$$S = \int_a^b \left(-mcd\tilde{s} - \frac{e}{c}A_i dx^i\right) \tag{43}$$

where the coordinates x^i are related to physical coordinates (t, x, y, z) by (26) and A_i are components of the (covariant) four-potential vector expressed through the contravariant components A^i by

$$(A_0, A_1, A_2, A_3) = \left(A^0, -A^1, -A^2, -A^3\right) \tag{44}$$

Upon representing the four-potential as

$$\left(A^0, A^1, A^2, A^3\right) = \left(\tilde{\phi}, \tilde{\boldsymbol{A}}\right) = \left(\tilde{\phi}, \tilde{A}_x, \tilde{A}_y, \tilde{A}_z\right) \tag{45}$$

where $A^0 = \tilde{\phi}$ is a scalar potential and the three-dimensional vector $\tilde{\boldsymbol{A}}$ is the vector potential of the field, the electromagnetic part of the action integral can be written in the form

$$S = \int_{\tilde{t}_1}^{\tilde{t}_2} \left(\frac{e}{c}\tilde{\boldsymbol{A}} \cdot \tilde{\boldsymbol{v}} - e\tilde{\phi}\right) d\tilde{t} \tag{46}$$

Here and in what follows, 'tilde' indicates that variables and operations are in Minkowskian space-time variables (26). Note that, while scalars and components of three-dimensional vectors in the Minkowskian formulation appear with 'tilde', four-dimensional Minkowskian variables are not supplied with 'tilde'. It does not lead to any confusion since the four-dimensional notation does not applicable to the formulation in physical variables.

In the electrodynamics of the standard special relativity (which, in our case, is electrodynamics in Minkowskian variables), the electric and magnetic field intensities are defined based on equations of motion of a charged particle obtained from the Lagrange equations.

$$\frac{d}{d\tilde{t}}\left(\frac{\partial \tilde{L}}{\partial \tilde{\boldsymbol{v}}}\right) = \frac{\partial \tilde{L}}{\partial \tilde{\boldsymbol{r}}} \tag{47}$$

where, in the Lagrange function $\tilde{L}$, a part related to the electromagnetic field is given by the integrand of (46). Then the electric and magnetic field intensities $\tilde{\boldsymbol{E}}$

and $\widetilde{\boldsymbol{H}}$ are introduced by separating the right-hand side of the vector equation of motion (the force) into two parts, one of which does not depend on the velocity of the particle and the second depends on the velocity, being proportional to the velocity and perpendicular to it, as follows.

$$\frac{d\tilde{\boldsymbol{p}}}{d\tilde{t}} = e\widetilde{\mathbf{E}} + \frac{e}{c}\tilde{\boldsymbol{v}} \times \widetilde{\boldsymbol{H}} \tag{48}$$

where $\widetilde{\boldsymbol{p}}$ is the momentum vector. The electric and magnetic field intensities are related to the potentials by

$$\widetilde{\boldsymbol{E}} = -\frac{1}{c}\frac{\partial \widetilde{\boldsymbol{A}}}{\partial \widetilde{t}} - \widetilde{\text{grad}}\ \widetilde{\phi}; \quad \widetilde{\boldsymbol{H}} = \widetilde{\text{curl}}\ \widetilde{\boldsymbol{A}} \tag{49}$$

The same line of arguments is used to derive equations describing the electromagnetic field in physical variables (t, x, y, z). The action integral is represented in the form

$$S = \int_{t_a}^{t_b} L dt \tag{50}$$

where t is the 'physical' time related to the Minkowskian variables via (26) and L is the Lagrangian in physical variables. The free particle part of L is defined by Eqs. (21)–(23). To obtain the electromagnetic field part of the Lagrangian, the right-hand side of (46) is transformed to physical space-time variables and then the new variables (ϕ, A_x, A_y, A_z) (modified potentials) are introduced by the relations

$$\begin{aligned} A^0 &= \widetilde{\phi} = \lambda(k)\phi, \ A^1 = \widetilde{A}_x = \lambda(k)(A_x - k\phi), \\ A^2 &= \widetilde{A}_y = \lambda(k)A_y, \ A^3 = \widetilde{A}_z = \lambda(k)A_z \end{aligned} \tag{51}$$

As the result, the Lagrangian function L in the action integral (50) takes the form

$$L = L_p + \frac{e}{c}\left(v_x A_x + v_y A_y + v_z A_z\right) - e\phi \tag{52}$$

where L_p is the free particle part of L defined by Eqs. (21)–(23). Substituting (52) into the Lagrange equations

$$\frac{d}{dt}\left(\frac{\partial L}{\partial \boldsymbol{v}}\right) = \frac{\partial L}{\partial \boldsymbol{r}} \tag{53}$$

yields

$$\frac{d\boldsymbol{p}}{dt} = -\frac{e}{c}\frac{\partial \boldsymbol{A}}{\partial t} - e\ \text{grad}\ \phi + \frac{e}{c}\boldsymbol{v} \times \text{curl}\ \mathbf{A} \tag{54}$$

Thus, upon using the modified potentials, equations of motion in physical variables have the same form as in the standard relativity and the physical electric and magnetic field intensities are expressed through the modified potentials by the relations

$$\mathbf{E} = -\frac{1}{c}\frac{\partial \boldsymbol{A}}{\partial t} - \text{grad}\ \phi; \quad \mathbf{H} = \text{curl}\ \mathbf{A} \tag{55}$$

of the same form (49) as in the standard relativity.

It is evident that the first pair of the Maxwell equations in physical variables, which is derived from Eq. (55), have the same form as in the standard relativity

$$\text{curl } \mathbf{E} = -\frac{1}{c}\frac{\partial \boldsymbol{H}}{\partial t}; \quad \text{div } \mathbf{H} = 0 \tag{56}$$

To obtain the second pair of Maxwell equations in physical variables let us calculate the components of the electromagnetic field tensor F_{ik} defined by

$$F_{ik} = \frac{\partial A_k}{\partial x^i} - \frac{\partial A_i}{\partial x^k} \tag{57}$$

Expressing A_i in the right-hand side of (57) through the modified potentials by (51) and then transforming the result to physical space-time variables using (26), with subsequent use of Eq. (55), yields the expressions for the components F_{ik} of the electromagnetic field tensor in terms of physical electric and magnetic field intensities. The result can be written as a matrix in which the index $i = 0, 1, 2, 3$ labels the rows, and the index k the columns, as follows

$$F_{ik} = \lambda(k)^2 \begin{pmatrix} 0 & E_x & E_y & E_z \\ -E_x & 0 & -H_z + kE_y & H_y + kE_z \\ -E_y & H_z - kE_y & 0 & -H_x \\ -E_z & -H_y - kE_z & H_x & 0 \end{pmatrix} \tag{58}$$

while

$$F^{ik} = \lambda(k)^2 \begin{pmatrix} 0 & -E_x & -E_y & -E_z \\ E_x & 0 & -H_z + kE_y & H_y + kE_z \\ E_y & H_z - kE_y & 0 & -H_x \\ E_z & -H_y - kE_z & H_x & 0 \end{pmatrix} \tag{59}$$

Note that the terms with k in the expressions (58) and (59) spoil the property, that $F^{ik} \to F_{ik}$ when $\boldsymbol{E} \to -\boldsymbol{E}$, of the standard relativity electrodynamics.

The electromagnetic field equations are obtained with the aid of the principle of least action [37] in the form

$$\frac{\partial F^{ik}}{\partial x^k} = 0 \tag{60}$$

(only fields in a vacuum, that are relevant to the subject of this paper, are considered). Substituting (59) into (60) and transforming the equations to physical space-time variables, upon combining equations with different 'i' and using the first pair of the Maxwell Eq. (56), yields the second pair of the Maxwell equations in the three-dimensional form

$$\text{div } \mathbf{E} = -\frac{k}{c}\frac{\partial E_x}{\partial t}; \quad \text{curl } \mathbf{H} = \left(1 - k^2\right)\frac{1}{c}\frac{\partial \boldsymbol{E}}{\partial t} - 2k\frac{\partial \boldsymbol{E}}{\partial x} + k \text{ grad } E_x \tag{61}$$

An important feature of Eq (61) is their linearity in $\boldsymbol{E}$ and $\boldsymbol{H}$ and hence in A^i. The Lorentz-violating terms thereby avoid the complications of nonlinear modifications to the Maxwell equations, which are known to occur in some physical situations such as nonlinear optics or when vacuum polarization effects are included. Another feature is that the extra Lorentz-violating terms involve only the electric field, as well as its derivatives.

Note the existence of an alternative way of the derivation of the modified Maxwell Eqs. (56) and (61). Based on Eqs. (49), (51), and (55), the electric and magnetic field intensities $\widetilde{\boldsymbol{E}}$ and $\widetilde{\boldsymbol{H}}$ in Minkowskian formulation can be related to the physical electric and magnetic field intensities $\boldsymbol{E}$ and $\boldsymbol{H}$, as follows

$$\begin{gathered}\widetilde{E_x} = \lambda(k)^2 E_x,\ \widetilde{E_y} = \lambda(k)^2 E_y,\ \widetilde{E_z} = \lambda(k)^2 E_z,\\ \widetilde{H_x} = \lambda(k)^2 H_x,\ \widetilde{H_y} = \lambda(k)^2 \left(H_y + kEz\right),\ \widetilde{H_z} = \lambda(k)^2 \left(H_z - kE_y\right)\end{gathered} \tag{62}$$

The same relations are seen in the expressions (58) for the components of the electromagnetic field tensor. It is readily verified that substituting the relations (62) into the Maxwell equations of the standard relativity

$$\widetilde{\text{curl}}\ \widetilde{\boldsymbol{E}} = -\frac{1}{c}\frac{\partial \widetilde{\boldsymbol{H}}}{\partial \widetilde{t}},\ \widetilde{\text{div}}\ \widetilde{\boldsymbol{H}} = 0,\quad \widetilde{\text{curl}}\ \widetilde{\boldsymbol{H}} = \frac{1}{c}\frac{\partial \widetilde{\boldsymbol{E}}}{\partial \widetilde{t}},\ \widetilde{\text{div}}\ \widetilde{\boldsymbol{E}} = 0 \tag{63}$$

as

$$\widetilde{E_x}(\tilde{t},\tilde{x},\tilde{y},\tilde{z}) = \lambda(k)^2 E_x(t(\tilde{t},\tilde{x}),x(\tilde{x}),y(\tilde{y}),z(\tilde{z})),\ \ldots \tag{64}$$

where

$$t(\tilde{t},\tilde{x}) = \lambda(k)\left(\tilde{t} + \frac{k}{c}\tilde{x}\right),\ x(\tilde{x}) = \lambda(k)\tilde{x}.,\ y(\tilde{y}) = \lambda(k)\tilde{y},\ z(\tilde{z}) = \lambda(k)\tilde{z} \tag{65}$$

yields the modified Maxwell Eqs. (56) and (61).

3.3 Electromagnetic waves

Like the electromagnetic wave equation of the standard relativity electrodynamics, the equation describing electromagnetic waves in the electrodynamics of the relativity with a preferred frame can be derived straight from the modified Maxwell equations (reproduced below for convenience)

$$\text{div}\ \mathbf{H} = 0,\quad \text{curl}\ \mathbf{E} = -\frac{1}{c}\frac{\partial \boldsymbol{H}}{\partial t} \tag{66}$$

$$\text{div}\ \mathbf{E} = -\frac{k}{c}\frac{\partial E_x}{\partial t};\quad \text{curl}\ \mathbf{H} = \left(1-k^2\right)\frac{1}{c}\frac{\partial \mathbf{E}}{\partial t} - 2k\frac{\partial \mathbf{E}}{\partial x} + k\ \text{grad}\ E_x \tag{67}$$

Eliminating $\boldsymbol{H}$ by taking 'curl' from the second equation of (66) and substituting curl $\mathbf{H}$ from the second equation of (67), with the subsequent use of differential consequences of the first equation of (67) for eliminating mixed space derivatives, yields

$$\frac{\partial^2 f}{\partial x^2} + \frac{\partial^2 f}{\partial y^2} + \frac{\partial^2 f}{\partial z^2} - \left(1-k^2\right)\frac{1}{c^2}\frac{\partial^2 f}{\partial t^2} + 2k\frac{1}{c}\frac{\partial^2 f}{\partial t\ \partial x} = 0 \tag{68}$$

where $f(t,x,y,z)$ stands for any component of $\boldsymbol{E}$. It is readily verified that the wave equation for $\boldsymbol{H}$ obtained from the modified Maxwell equations in a similar way has the same form (68).

Alternatively, the wave Eq. (68) can be derived from (60) expressed in terms of the potentials using (57) while imposing the Lorentz gauge condition

$$\frac{\partial A^k}{\partial x^k} = 0 \tag{69}$$

Converting the derivatives in the resulting equation

$$\frac{\partial^2 A^k}{\partial x_k\ \partial x^k} = 0 \tag{70}$$

into derivatives in physical space-time variables yields equations of the form (68) with f being any component of A^k. Given the fact, that equations (51) and (55) relating A^k to the modified potentials $(\phi, \boldsymbol{A})$ and then to $\boldsymbol{E}$ and $\boldsymbol{H}$ are linear, it is evident that any of those variables obeys Eq. (68).

Much of the propagation behavior of the electromagnetic wave is encoded in its dispersion relation, which provides spectral information for the modes. To find the dispersion relation the ansatz in the form of monochromatic plane waves is used, as follows

$$f(t,x,y,z) = f_a(\omega, \boldsymbol{q})) exp \left[i \left(q_x x + q_y y + q_z z - \omega t\right)\right] \tag{71}$$

where ω and $\boldsymbol{q} = \left(q_x, q_y, q_z\right)$ can be regarded as the frequency and wave vector of the mode or as the associated energy and momentum (taking the real part is understood, as usual). Substituting (71) into (68) yields the dispersion relation

$$c^2q^2 - 2ckq_x\omega - \left(1-k^2\right)\omega^2 = 0 \quad \text{where } q^2 = q_x^2 + q_y^2 + q_z^2 \tag{72}$$

The dispersion relation (72) can be also represented in the form

$$\left(\frac{\omega}{c^{(+)}} - q_x\right)\left(\frac{\omega}{c^{(-)}} + q_x\right) = q_y^2 + q_z^2 \tag{73}$$

where $c^{(+)}$ and $c^{(-)}$ are defined by (2). The form (73) adheres to the dispersion relation (40) for free massless particles with E and $\boldsymbol{P}$ replaced by ω and $\boldsymbol{q}$. In the standard relativity, the polynomial (72) determining ω reduces to one with two quadruply degenerate roots $\omega = \pm cq$ which correspond to the opposite directions of the group velocity. In the modified electrodynamics, the polynomial also has two roots

$$\omega = c\frac{-kq_x + \sqrt{\left(1-k^2\right)q^2 + k^2q_x^2}}{1-k^2}, \qquad \omega = c\frac{-kq_x - \sqrt{\left(1-k^2\right)q^2 + k^2q_x^2}}{1-k^2} \tag{74}$$

Like as in the standard relativity case, the two roots (74) are obtained from each other by changing the sign of ω but, in the case of $k \neq 0$, it is accompanied by a change of sign of the anisotropy parameter k.

More insight about the wave motion implied by Eq. (68) can be gained from the modified Maxwell Eqs. (66) and (67). Eq. (66), which are unaffected by the modifications, reduce with the ansatz (71) to

$$\boldsymbol{q} \cdot \boldsymbol{H} = 0, \quad \omega\, \boldsymbol{H} = -\boldsymbol{q} \times \boldsymbol{E} \tag{75}$$

The first of these equations shows that the magnetic field remains transverse to $\boldsymbol{q}$ despite the Lorentz violation. The second equation shows that the magnetic field $\boldsymbol{H}$ is perpendicular to the electric field $\boldsymbol{E}$. The first equation of (67) reduces to

$$\boldsymbol{q} \cdot \boldsymbol{E} = \omega \frac{k}{c} E_x \tag{76}$$

Eq. (76) implies the existence of two modes.

The first one corresponds to the electric field with $E_x = 0$. Then it follows from (76) that the electric field is perpendicular to $\boldsymbol{q}$. Further, the condition $E_x = 0$ implies that the vector $\boldsymbol{E}$ lies in the plane (y, z) and so the vector $\boldsymbol{q}$ is directed along the x -axis (the direction of the anisotropy vector $\boldsymbol{k}$). Therefore $H_x = 0$ and also, based on rotational symmetry in the plane (y, z), it can be set $H_z = 0$ which implies $E_y = 0$. In such a case, the first equation of (66) shows that $q_y = 0$ and the first equation of (66) shows that $q_z = 0$. Then the second equation of (66) and the second equation of (67) reduce to the system of equations for the two nonzero components of the electric and magnetic field intensities E_z and H_y while the requirement of vanishing the determinant of the system yields the dispersion relation (72). Thus, the mode with $E_x = 0$ represents a usual electromagnetic plane wave with the magnetic and electric fields transverse to the direction of propagation of the wave $\boldsymbol{q}$ and perpendicular to each other, which propagates along the direction of the anisotropy vector (but with the modified dispersion relation).

The second mode corresponds to the case $E_x \neq 0$. Then it follows from (76) that the electric field vector is not normal to $\boldsymbol{q}$. Since, according to the second equation of (75), $\boldsymbol{H}$ is normal to the plane of $\boldsymbol{E}$ and $\boldsymbol{q}$, one can choose, without loosing generality, the direction of $\boldsymbol{H}$ to be along the y -axis and the plane of the vectors $\boldsymbol{q}$ and $\boldsymbol{E}$ to be the (x, z) -plane. Then the first equation of (75) gives $q_y = 0$ and it is readily verified that the remaining equations of (66) and (67) can be satisfied only if $q_z \neq 0$ with ω, q_x and q_z obeying the dispersion relation (72) where it is set $q_y = 0$. Note the particular case, when $\boldsymbol{E}$ is directed along the x -axis ($E_z = 0$), in which the dispersion relation degenerates to

$$\omega = \frac{cq_x}{k}, \quad q_z = \pm \frac{q_x}{k} \tag{77}$$

Thus, the second mode represents electromagnetic wave, in which the magnetic field $\boldsymbol{H}$ is transverse to direction of propagation $\boldsymbol{q}$ and perpendicular to the electric field $\boldsymbol{E}$, like as in the regular wave, but, as distinct from the regular wave, the electric field is not normal to $\boldsymbol{q}$. Another characteristic feature of such a wave, that distinguishes it from the first mode, is that the direction of propagation is not along the anisotropy vector $\boldsymbol{k}$ and so not along with the velocity of relative motion of the source and the observer. It implies that in the case when the relative motion velocity is only the cosmological recession velocity, such a wave propagates not along a line of sight.

It is worthwhile to note a distinguishing feature of the above analysis as compared with other studies of electromagnetic waves in the presence of the Lorentz violation. Typically, different modes arising due to the Lorentz violation correspond to different roots of the modified dispersion relation (see, e.g., [6, 38–40]). The present analysis provides an unusual example when two different modes correspond to the same root of the dispersion relation (for the waves propagating to the observer. it is the second root of (74)). The existence of two modes is revealed only when one studies the corresponding solutions of the modified Maxwell equations. It is worth also noting that the present analysis is performed solely in terms of field intensities $\boldsymbol{E}$ and $\boldsymbol{H}$ while most studies of electromagnetic waves in the presence of the Lorentz violation involve also the electromagnetic field potentials A^k which are accompanied by extensive discussions of different gauge choices and their influence on the results.

4. Cosmology

4.1 General relativity

The basic principle of general relativity, the Equivalence Principle (see, e.g. [41]), which asserts that at each point of spacetime it is possible to choose a 'locally inertial' coordinate system where objects obey Newton's first law, is valid independently of the law of propagation of light assumed. In other terms, it can be applied when the processes in the locally inertial frame are governed by the laws of 'relativity with a preferred frame'. Based on that there exists the invariant combination (7), which by the change of variables (9) is converted into the Minkowski interval, one can state that the general relativity equations in arbitrary coordinates (x^0, x^1, x^2, x^3) are valid if the locally inertial coordinates $(\xi^0, \xi^1, \xi^2, \xi^3)$ are

$$\xi^0 = c\tilde{t}, \; \xi^1 = \tilde{x}, \; \xi^2 = \tilde{y}, \; \xi^3 = \tilde{z} \tag{78}$$

where $\tilde{t}, \tilde{x}, \tilde{y}$ and $\tilde{z}$ are defined by (9). In these variables, the invariant spacetime distance squared $ds^2 = g_{ik}dx^i dx^k$ is equal to $d\tilde{s}^2 = \eta_{ik}d\xi^i d\xi^k$ (the notation η_{ik} is used for the Minkowski metric and the rule of summation over repeated indices is implied). Thus, the apparatus of general relativity is applied in the coordinates (x^0, x^1, x^2, x^3) while, in the calculation of the 'true' time and space intervals, the 'physical' variables (t^*, x^*, y^*, z^*) (it is the new notation for what was before (t, x, y, z)) are to be used. Eq. (9) relating the physical coordinates to the 'locally inertial' coordinates, rewritten with allowance for (78) and (9), are

$$t^* = \frac{1}{c}\lambda(k)\left(\xi^0 + k\xi^1\right), \; x^* = \lambda(k)\xi^1, \; y^* = \lambda(k)\xi^2, \; z^* = \lambda(k)\xi^3 \tag{79}$$

The 'true' time and space intervals can be determined using a procedure similar to that described in [37]. Applying that procedure (see [27] for details) yields the following relations for the 'true' proper time interval dt^* and the element dl^* of 'the true' spatial distance:

$$dt^* = \frac{1}{c}\lambda(k)\sqrt{g_{00}}dx^0 \tag{80}$$

$$dl^* = \lambda(k)\sqrt{\gamma_{\alpha\beta}dx^\alpha dx^\beta}, \quad \gamma_{\alpha\beta} = -g_{\alpha\beta} + \frac{g_{0\alpha}g_{0\beta}}{g_{00}} \tag{81}$$

where g_{ik} $(i, k = 0, 1, 2, 3)$ are components of the space-time metrical tensor and $\gamma_{\alpha\beta}$ $(\alpha, \beta = 1, 2, 3)$ are components of the space metrical tensor. It is important to note, that the expression for the proper velocity of a particle $v = dl^*/dt^*$ is not modified, since the time and the distance intervals are modified by the same factor $\lambda(k)$.

4.2 Cosmological models

Modern cosmological models assume that, at each point of the universe, the 'typical' (freely falling) observer can define the (preferred) Lorentzian frame in which the universe appears isotropic. The metric derived based on isotropy and homogeneity (the *Robertson-Walker metric*) has the form [41, 42]

$$ds^2 = dt^2 - a^2(t)\left(\frac{dr^2}{1 - K_c r^2} + r^2 d\Omega\right), \quad d\Omega = d\theta^2 + \sin^2\theta d\phi^2 \tag{82}$$

where a comoving reference system, moving at each point of space along with the matter located at that point, is used. This implies that the coordinates (r, θ, ϕ) are unchanged for each typical observer. In (82), and further throughout this section, the system of units in which the speed of light is equal to unity, is used. The time coordinate $x^0 = t$ is the synchronous proper time at each point of space. The constant K_c (this notation is used, instead of common k or K, to avoid confusion with the symbols for the anisotropy parameter) by a suitable choice of units for r can be chosen to have the value $+1$, 0, or -1.

Introducing, instead of r, the radial coordinate χ by the relation $r = S(\chi)$ with

$$S(\chi) = \begin{cases} \sin\chi & \text{for } K_c = 1 \\ \sinh\chi & \text{for } K_c = -1 \\ \chi & \text{for } K_c = 0 \end{cases} \tag{83}$$

and replacing the time t by the *conformal time* η defined by

$$dt = a(t)d\eta \tag{84}$$

converts (82) into the form

$$ds^2 = a^2(\eta)\left[d\eta^2 - d\chi^2 - S^2(\chi)\, d\Omega\right] \tag{85}$$

The information about the scale factor $a(t)$ in the Robertson-Walker metric can be obtained from observations of shifts in the frequency of light emitted by distant sources. The frequency shift can be calculated by considering the propagation of a light ray in isotropic space with the metric (85) adopting a coordinate system in which we are at the center of coordinates $\chi = 0$ and the source is at the point with a coordinate $\chi = \chi_1$. A light ray propagating along the radial direction obeys the equation $d\eta^2 - d\chi^2 = 0$. For a light ray coming toward the origin from the source, that equation gives

$$\chi_1 = -\eta_1 + \eta_0 \tag{86}$$

where η_1 corresponds to the moment of emission t_1 and η_0 corresponds to the moment of observation t_0. The *red-shift parameter* z is defined by

$$z = \frac{\nu_1}{\nu_0} - 1 \tag{87}$$

where ν_0 is the observed frequency and ν_1 is the frequency of the emitted light which coincides with the frequency of a spectral line observed in terrestrial laboratories. Calculations within the framework of the relativity with a preferred frame (see details in [27]) lead to the relation

$$z = \frac{a(\eta_0)}{a(\eta_0 - \chi_1)B(\bar{\beta}_1)} - 1 \tag{88}$$

The relation expressing the *Luminosity Distance* d_L of a cosmological source in terms of its redshift z is one of the fundamental relations in cosmology. It has been exploited to get information about the time evolution of the expansion rate. In a matter-dominated cosmological model of the universe (Friedman-Robertson-Walker model) based on the standard GR, solving the gravitational field equations yields the luminosity distance-redshift relation of the form

$$d_L = H_0^{-1}\left(z + \frac{1}{2}\left(1 - q_0^{(D)}\right)z^2 + \cdots\right) \tag{89}$$

where the deceleration parameter $q_0^{(D)}$ is positive for all three possible values of the curvature parameter K_c which means that, in that model, the expansion of the universe is decelerating. However, recent observations of Type Ia supernovae (SNIa), fitted into the luminosity distance versus redshift relation of the form (89), corresponding to the deceleration parameter $q_0^{(D)} < 0$ which indicates that the expansion of the universe is accelerating. This result is interpreted as that the time evolution of the expansion rate cannot be described by a matter-dominated cosmological model. To explain the discrepancy within the context of general relativity and fit the theory to the SNIa data, the dark energy, a new component of the energy density with strongly negative pressure that makes the universe accelerate, is introduced (see, e.g., [42]).

In the relativity with a preferred frame, solving the modified GR equations for a matter-dominated model lead to the luminosity distance-redshift relation of the form, which allows fitting the results of observations with supernovae so that the acceleration problem can be naturally resolved—there is no acceleration and so no need in introducing the dark energy. Below, the calculations leading to the modified luminosity distance-redshift relation are outlined (for more details see [27]).

In the relativity with a preferred frame, the expression for d_L is obtained in the form [27]

$$d_L = a(\eta_0)(1+z)S(\chi_1) \tag{90}$$

which coincides with a common form of the relation for d_L [37, 42]. Nevertheless, even though it does not contain the factor $B(\bar{\beta}_1)$, the dependence of d_L on z obtained by eliminating χ_1 from Eqs. (90) and (88) will differ from the common one since the relation (88) for z does contain the factor $B(\bar{\beta}_1)$. To derive the dependence $d_L(z)$ in a closed-form using Eqs. (90) and (88), the function $a(\eta)$ determining the dynamics of the cosmological expansion it to be defined by solving the gravitational field equations of Einstein which requires to make some tentative assumptions about the cosmic energy density ρ and the form of equation of state giving the pressure p as a function of the energy density. The energy density $\rho(t)$ is usually assumed to be a mixture of non-relativistic matter with equation of state $p = 0$ and dark energy with equation of state $p = w\rho$ while ignoring the relativistic matter (radiation). In the commonly accepted ΛCDM model, the dark energy obeys the equation of state with $w = -1$ (vacuum) which is equivalent to introducing into Einstein's equation a cosmological constant Λ. Then the fundamental Friedmann equation, which is obtained as a consequence of the Einstein field equations, can be written in the form (see, e.g., [42])

$$(x')^2 = H_0^2 x^2\left(\Omega_\Lambda + \Omega_M x^{-3} + \Omega_K x^{-2}\right) \tag{91}$$

where

$$x(t) = \frac{a(t)}{a_0}, \quad a_0 = a(t_0) \tag{92}$$

and the parameters Ω_Λ, Ω_M and Ω_K are defined by

$$\Omega_\Lambda = \frac{\rho_{V0}}{\rho_c}, \quad \Omega_M = \frac{\rho_{M0}}{\rho_c}; \qquad \rho_c = \frac{3H_0^2}{8\pi G}, \quad \Omega_K = -\frac{K_c}{a_0^2 H_0^2} \tag{93}$$

where G is Newton's gravitational constant, ρ_{V0} and ρ_{M0} are the present energy densities in the vacuum and non-relativistic matter and ρ_c is the critical energy density. Being evaluated at $t = t_0$ Eq. (91) becomes

$$\Omega_\Lambda + \Omega_M + \Omega_K = 1 \tag{94}$$

The Friedmann Eq. (91) allows us to calculate the radial coordinate χ_1 of an object of a given redshift z. Eq. (86) defining χ_1 can be represented in the form

$$\chi_1 = \eta_0 - \eta_1 = \int_{\eta_1}^{\eta_0} d\eta = \int_{t_1}^{t_0} \frac{dt}{a(t)} = \frac{1}{a_0} \int_{x_1}^{1} \frac{dx}{x'x} \tag{95}$$

where x' is a function of x defined by the Friedmann Eq. (91) and $x_1 = a(t_1)/a_0$. Then using Eq. (91) in (95) yields

$$\chi_1 = \int_{x_1}^{1} \frac{dx}{a_0 H_0 x^2 \sqrt{\Omega_\Lambda + \Omega_M x^{-3} + \Omega_K x^{-2}}} \tag{96}$$

In the standard cosmology, Eq. (88) (with $B(\bar{\beta}_1) = 1$) provides a simple relation

$$x_1 = \frac{1}{1+z} \tag{97}$$

so that (96) becomes a closed-form relation for $\chi_1(z)$. For a 'concordance' model, which is the flat space ΛCDM model, $\Omega_K = 0$ and $\Omega_\Lambda = 1 - \Omega_M$, Eq. (96) can be represented in the form

$$\chi_{1m}^c(z_1) = \int_0^{z_1} \frac{dz}{\sqrt{1 - \Omega_M + \Omega_M(1+z)^3}}, \quad \chi_{1m}^c = \chi_1^c a_0 H_0 \tag{98}$$

Here and in what follows, quantities with a superscript "c" refer to the concordance model, with the original notation secured for the corresponding quantities of the present model. Then the luminosity distance is calculated as

$$d_L^c(z_1) = \frac{1}{H_0}(1+z_1)\chi_{1m}^c(z_1) \tag{99}$$

In the framework of the present analysis, expressing χ_1 as a function of z_1 by combining Eqs. (96) and (88) becomes more complicated in view of the fact that $\bar{\beta}_1$, and so the factor $B(\bar{\beta}_1)$, depend on χ_1. We will outline the calculations for the case of a flat universe, $\Omega_K = 0$, which is also the assumption of the concordance model.[¶] With that assumption and the presumption, that in the cosmology based on the relativity with a preferred frame there is no need in introducing dark energy,

¶ In the present model, this assumption is not obligatory. It is worthwhile to note that, despite what is frequently claimed, a flatness of the universe is not stated in modern cosmology. Given the fact, that there is no direct measurement procedure of the curvature of space independent of the cosmological model assumed, the flatness of the space is the result valid only within the framework of the ΛCDM model.

$\Omega_\Lambda = 0$, the relation $\chi_1(x_1)$ is obtained from (96) in an analytical form, which allows finding the dependence $x_1(\chi_1)$ by inverting the result, as follows

$$\chi_{1m} = 2(1 - \sqrt{x_1}) \quad \Rightarrow \quad x1 = \frac{1}{4}(\chi_{1m} - 2)^2, \quad \chi_{1m} = a_0 H_0 \chi_1 \tag{100}$$

The dependence $B(\chi_1)$, with the accuracy up to third order in χ_1, is given by [27]

$$B(\bar{\beta}_1(\chi_1)) = 1 - \frac{b}{2}\left(\chi_{1m}^2 + \Omega_M \chi_{1m}^3\right) \tag{101}$$

Substituting (101) and (100) into (88) reduces the problem to a transcendental equation for $\chi_{1m}(z_1)$, as follows

$$\frac{1}{4}(\chi_{1m}(z_1) - 2)^2(z_1 + 1)\left(1 - \frac{b}{2}\left(\chi_{1m}(z_1)^2 + \chi_{1m}(z_1)^3\right)\right) = 1 \tag{102}$$

Representing the solution of (102) as a series in z_1 yields

$$\chi_{1m}(z_1) = z_1 + \frac{1}{4}(-3 - 2b)z_1^2 + \frac{1}{8}(5 + 4b + 4b^2)z_1^3 \tag{103}$$

Then the relation $d_L(z_1)$, calculated from (90) with $S(\chi_1) = \chi_1$, is

$$d_L(z_1) = \frac{1}{H_0}\left(z_1 + \frac{1}{4}(1 - 2b)z_1^2 + \frac{1}{8}(4b^2 - 1)z_1^3\right) \tag{104}$$

To compare the results produced by the model with those, obtained from an analysis of type Ia supernova (SNIa) observations, one needs some fitting formulas for the dependence $d_L(z)$ derived from the observational data. It is now common, in an analysis of the SNIa data, to fit the Hubble diagram of supernovae measurements to the ΛCDM model (mostly, to the concordance model) and represent the results as constraints on the model parameters (see, e.g. [43]). Therefore, in what follows, a comparison of the results with the SNIa data is made by comparing the dependence $d_L(z)$ produced by the present model with $d_L^c(z)$ for the concordance model while using constraints on the parameter Ω_M^c from the SNIa data analysis. It is found that, for every value of Ω_M^c from the interval, defined by fitting the SNIa data to the concordance model, the parameter b can be chosen such that the dependence $d_L(z)$ coincided with $d_L^c(z)$ with a quite high accuracy (were graphically undistinguishable). An example is given in **Figure 1** where the dependence $d_L(z)$ for $\Omega_M = 1$ (flat universe), defined by Eq. (104), is plotted for three different values of b together with $d_L^c(z)$ of the concordance model. It demonstrates that there exists a value of b (in the present case it is $b = 0.672$) for which the deviation is negligible. As it was mentioned above, in the present model the assumption of the flat universe is not obligatory. Calculations for other values of Ω_M (remind that $\Omega_K = 1 - \Omega_M$) show that for every value of $\Omega_M > 0$ there exists the value of b, for which the deviation $d_L(z)$ from $d_L^c(z)$ is negligible. It is worth clarifying again that the above is intended to be a comparison of the dependence $d_L(z)$ yielded by the present model with that derived from the SNIa observations so that the dependence $d_L^c(z)$ for the 'concordance' model plays a role of a fitting formula for the SNIa data.

The Baryon Acoustic Oscillations (BAO) data are commonly considered as confirming the accelerated expansion and imposing constraints on the dark energy parameters. Applying the cosmological models based on the 'relativity with a preferred frame' to the interpretation of the BAO data provides an alternative view on

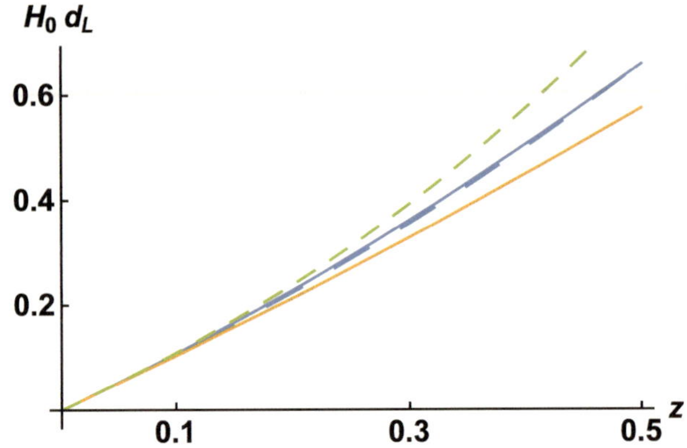

Figure 1.
Dependence of the luminosity distance d_L on the red-shift z: thin solid line for the concordance model with $\Omega_M^c = 0.31$; short-dashed for the present model with $\Omega_M = 1$, $b = -1.2$; long-dashed for the present model with $\Omega_M = 1$, $b = -0.672$; thick solid for the present model with $\Omega_M = 1$, $b = -0.2$.

the role of the BAO observations in cosmology. Comparing the predictions of the present model with the recently released galaxy clustering data set of the Baryon Oscillation Spectroscopic Survey (BOSS), part of the Sloan Digital Sky Survey III (SDSS III), shows that the BAO data can be well fit to the present cosmological model. The BAO data include two independent sets of data: the BAO scales in transverse and line-of-sight directions which can be interpreted to yield the comoving angular diameter distance $D_M(z)$ and the Hubble parameter $H(z)$ respectively. In [44], the results of several studies studying the sample provided by the BOSS data with a variety of methods are combined into a set of the final consensus constraints on $D_M(z)$ and $H(z)$ that optimally capture all of the information. It is found (see [27] for details) that the results yielded by the present model are consistent with the consensus constraints of [44] on both $D_M(z)$ and $H(z)$. The two regions in the plane (Ω_M, b) defined by constraints on these two sets are overlapped such that the overlapping area corresponds to the values of the model parameters for which the results on $H(z)$ and $D_M(z)$ are consistent both with the BAO data and with each other. And what can be considered as a very convincing proof of the robustness of the present model is that a line in the plane $(\Omega_M.b)$, on which the results produced by the present model fit also the SNIa observational data, passes inside that quite narrow overlapping region defined by the BAO data. Thus, the results produced by the present model fit three different sets of data by adjusting (together with the matter density parameter Ω_M) only one universal parameter b. It is worth noting again that, as distinct from the concordance model to which the SNIa and BAO data are commonly fitted by adjusting the dark energy parameters, the present model fits well all the data without introducing dark energy.

5. Propagation of cosmic rays

5.1 Attenuation of the UHECR due to the pion photoproduction process

In this section, the application of the theory to the description of the effects due to the interactions of the Ultra-High Energy Cosmic Rays (UHECR) with universal diffuse background radiation in the course of the propagation of cosmic rays from their sources to Earth over long distances (see, e.g., review articles [45–47]) is considered. The interactions of the UHECR with the CMB photons are characterized by a well-defined energy threshold for the energy suppression due to pion photoproduction by UHECR protons—the Greisen-Zatsepin-Kuzmin (GZK) cutoff [48, 49]. The fluxes of cosmic ray protons with energies above this threshold would be strongly attenuated over distances of a few tens of Mpc so that the cosmic ray protons from the sources at a larger distance, even if they were accelerated to energies higher than the threshold, would not be able to survive the propagation. The energy position of the GZK cutoff can be predicted based on special relativity as a theoretical upper limit ('GZK limit') on the energy of UHECR set by pion photoproduction in the interactions of cosmic ray particles with the microwave background radiation. Calculating the GZK limit based on the particle dynamics of the special relativity with a preferred frame developed in Section 3.1 (see [28] for details) yields

$$\frac{Eth}{Est} = \left(1 - z^2\right)^{-b}; \quad Est = \frac{\varepsilon_\pi\left(2\varepsilon_p + \varepsilon_\pi\right)}{4E^{(\gamma)}} \tag{105}$$

where Eth is the threshold value of the UHECR protons energy calculated using equations of the relativity with a preferred frame, Est is the standard value of the GZK threshold calculated using equations of the standard relativity, $\varepsilon_p = m_p c^2$ and $\varepsilon_\pi = m_\pi c^2$ are the proton and pion rest energies and $E^{(\gamma)}$ is the CMB photon energy.

It is seen that the expression (105) for the threshold energy of the proton differs from the common one by the factor $\left(1 - z^2\right)^{-b}$. The universal constant b is negative, both as it is expected from intuitive arguments and as it is found by fitting the cosmological model developed in the framework of the 'relativity with a preferred frame' to the observational data (Section 4.2). Therefore the threshold energy decreases as the distance to the source of the particles (the redshift z) increases (**Figure 2**, left panel).

This effect may contribute to the interpretation of the data on the mass composition of UHECR which is a key observable in the context of the physics of UHECR as it fixes few fundamental characteristics of the sources. The mass composition of UHECR became a matter of active debate after that the Pierre Auger Collaboration (Auger) reported on its recent observations [50, 51]. The observations of Auger, far the largest experiment set-up devoted to the detection of UHECR, have shown that the UHECR mass composition is dominated by protons only at energies around and below 10^{18} eV and then the fraction of protons is progressively decreasing up to

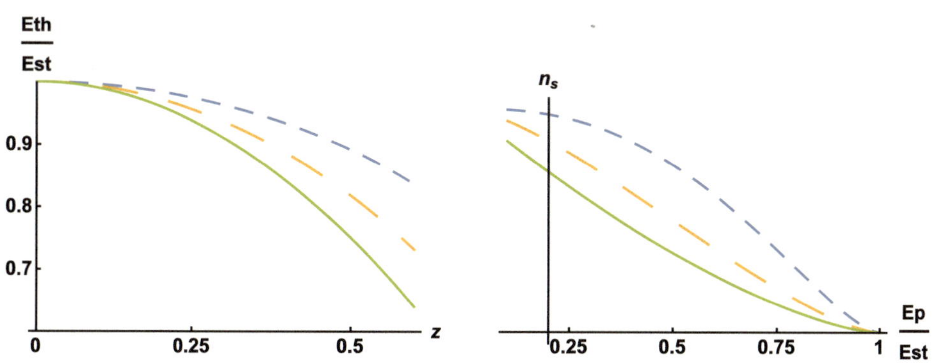

Figure 2.
Left panel: dependence of the correction factor to the GZK threshold on the source redshift z for different values of the parameter b: shot-dashed for b = −0.4; long-dashed for b = −0.7; solid for b = −1. Right panel: Number of sources n_s (in arbitrary units), that may contribute to the observed flux of protons at the energy Ep, versus $\frac{Ep}{Est}$, where Est is the standard GZK threshold value, for different values of b: shot-dashed for b = −0.4; long-dashed for b = −0.7; solid for b = −1.

energies of $10^{19.6}$ eV. It seemed to be not consistent with the general consensus, that UHECRs are mostly protons and that sources should accelerate them to $> 10^{20}$ eV. At the same time, the Telescope Array (TA) experiment, even if with 1/10 of the Auger statistics, collected data seemed to confirm the pre-Auger scenario [52]. A common effort of the Auger and TA collaborations allowed to reconcile the interpretations of the Auger and TA observations so that the evidence for a composition becoming gradually heavier towards higher energies is now considered to be well established. It implies that the primary UHECR flux at the sources includes both protons and heavy nuclei which are to be accelerated with very high maximum injection energies. This imposes severe constraints on the parameters of the acceleration models and has served as a stimulus to build new acceleration models or reanimate the previously developed models that can potentially explain the phenomenology of the UHECR mass composition data. The models are characterized by a complex scenario and/or include some exotic assumptions.

The complexity of the scenario and the severe constraints on the model parameters, required in the case of a composition with heavy nuclei, are not present in the case if the UHECR mass composition is dominated by protons. In the latter case, the scenario is much simpler, only protons are accelerated with very high maximum injection energies. The view that the UHECR are mostly protons is, theoretically, a natural possibility. Proton is the most abundant element in the universe and several different astrophysical objects, at present and past cosmological epochs could provide efficient acceleration even if it requires very high luminosities and maximum acceleration energies. The models of interaction of UHECR with the astrophysical background are also much simpler if the UHECR are mostly protons. In this case, the only relevant astrophysical background is the CMB [53, 54]. This fact makes the propagation of UHE protons free from the uncertainties related to the background, being the CMB exactly known as a pure black body spectrum that evolves with red-shift through its temperature.

The results of the present study allow reconciling (at least, partially) the view, that, the primary UHECR flux at the sources is dominated by protons accelerated with very high maximum injection energies, with the observational evidence that the fraction of protons in the UHECR is decreasing towards higher energies. The apparent contradiction can be resolved by taking into account the effect, predicted by the present analysis, that the number of sources, which may contribute to the observed flux of protons at a given energy, is progressively decreasing with the energy increases. This effect is a consequence of the threshold condition (105) which implies that, among protons produced by a source at some z, only those having the energies lower than the threshold energy for that z, can reach the Earth. In other terms, for a given value Ep of the proton energy, there exists a value z_{th} of the redshift (distance D_{th}) such that, for the UHECR sources with $D > D_{th}$, the GZK threshold Eth is less than Ep and so the protons with the energy Ep injected by the sources at the distances $D > D_{th}$ cannot reach the Earth. Thus, the sources, that may contribute to the observed flux at the energy Ep, are confined within the sphere of the radius D_{th}, with D_{th} decreasing when the proton energy Ep is increasing. If the distribution of the sources in space is more or less uniform, the number of sources Ns, that may contribute to the observed flux at the energy Ep, decreases with Ep (**Figure 2**, right panel). Thus, reducing the fraction of protons in the observed UHECR flux towards the higher energies can be considered as the result of reducing the number of sources contributing to the flux.

5.2 Attenuation due to the pair-production process

Gamma rays (γ) propagating from distant sources to Earth interact with the photons of the extragalactic background light (γ_b) being able to produce e^+e^- through the process of pair production

$$\gamma + \gamma_b \rightarrow e^+ + e^- \tag{106}$$

which has the effect of a significant energy attenuation in the flux of high-energy gamma rays. Such interaction takes place for gamma rays with energies (E_γ) above the threshold of pair production. The existence of a threshold can be also expressed as the minimum energy ($E^{th}_{\gamma_b}$) that a γ_b needs to produce a e^+e^-.

The following assumptions should be made if we intend to calculate the threshold value of the energy of the gamma-rays photons:

i. It is needed to take the lowest energy the high-energy photon can have to react with the background photon to yield the two particles which correspond to the situation when they both are produced at rest in their center of mass frame after the collision.

ii. To maximize the energy available from the collision, the initial momenta of the two particles in the lab frame should be pointing in opposite directions.

Let us equate the square of the total 4—momentum $p^{(L)} = p^{(\gamma)} + p^{(\gamma_b)}$ in the lab frame before the collision with the square of the total 4—the momentum of the outgoing particles $p^{(CM)} = p^{(+)} + p^{(-)}$ in their center of mass frame after the collision

$$\left(p^{(\gamma)} + p^{(\gamma_b)}\right)^2 = \left(p^{(+)} + p^{(-)}\right)^2 \tag{107}$$

The right-hand side of (107) is calculated, as follows

$$\left(p^{(+)} + p^{(-)}\right)^2 = \left(p_0^{(+)} + p_0^{(-)}\right)^2 - \left(p_1^{(+)} + p_1^{(-)}\right)^2 - \left(p_2^{(+)} + p_2^{(-)}\right)^2 - \left(p_3^{(+)} + p_3^{(-)}\right)^2 \tag{108}$$

where Eq. (39) are to be substituted into (108), with the three-momentum and energy defined by equations (24), (25) and (21) in which it is set $\beta_x = \beta_y = \beta_z = 0$ for both particles. As the result, we obtain the following expression for the square of the total 4—momentum of outgoing particles

$$\left(p^{(+)} + p^{(-)}\right)^2 = c^2(m_e + m_e)^2 \tag{109}$$

Note that, although P_x does not vanish for $\beta_x = \beta_y = \beta_z = 0$, the component p_1 of the four-momentum does vanish since, in the expression (39) for p_1, the first term compensates the non-vanishing part of P_x.

The left-hand side of Eq. (107) is to be expressed in terms of the high-energy and background photons energies using the relations between the particle's momentum and energies obtained from the dispersion relation (40). The high-energy photons move to the observer, in the direction opposite to the direction the velocity of the lab frame relative to the observer (relative to the preferred frame) which is chosen to be a positive direction of the x-axis. So, the high-energy photon moves along the x-axis in the negative x direction while the background photon moves, according to the threshold assumption (ii), in the positive x direction. Thus, the momenta of the photons are related to their energies using Eq. (41), as follows

$$P_x^{(\gamma)} = -\frac{E_\gamma(1-k)}{c}, \quad P_x^{(\gamma_b)} = \frac{E_{\gamma_b}(1+k)}{c} \tag{110}$$

where k is the anisotropy parameter in the lab frame. Then the left-hand side of (107) is calculated as follows (head-on collision)

$$\left(p^{(\gamma)}+p^{(\gamma_b)}\right)^2=\left(p_0^{(\gamma)}+p_0^{(\gamma_b)}\right)^2-\left(p_1^{(\gamma)}+p_1^{(\gamma_b)}\right)^2$$
$$=\left(\frac{E_\gamma\lambda(k)}{c}+\frac{E_{\gamma_b}\lambda(k)}{c}\right)^2-\left(\lambda(k)\left(\frac{kE_\gamma}{c}-P_x^{(\gamma)}\right)+\lambda(k)\left(\frac{kE_{\gamma_b}}{c}-P_x^{(\gamma_b)}\right)\right)^2 \quad (111)$$

Substituting (110) for $P_x^{(\gamma)}$ and $P_x^{(\gamma_b)}$ into (111) yields

$$\left(p^{(\gamma)}+p^{(\gamma_b)}\right)^2=4\lambda(k)^2\frac{E_\gamma E_{\gamma_b}}{c^2} \quad (112)$$

Then using Eqs. (112) and (109) in (107) and solving the resulting equation for E_γ one obtains the expression for the threshold energy of the high-energy photon

$$E_\gamma^{th}=\frac{m_e^2c^4}{\lambda(k)^2E_{\gamma_b}} \quad (113)$$

or the expression for the threshold energy of the background photon (minimum energy to produce e^+e^-)

$$E_{\gamma_b}^{th}=\frac{m_e^2c^4}{\lambda(k)^2E_\gamma} \quad (114)$$

The factor $\lambda(k)$ can be represented as a function $B(\bar{\beta})$ of the frame velocity $\bar{\beta}$ relative to a preferred frame which, with an accuracy up to $(\bar{\beta})^3$, is given by the expression (see (17))

$$B(\bar{\beta})=\left(1-\bar{\beta}^2\right)^{b/2} \quad (115)$$

In a cosmological context, where $\bar{\beta}$ is a recession velocity of a source, $\bar{\beta}$ depends on the cosmological redshift of an object z. Although the expansion of $\bar{\beta}(z)$ in series, besides the leading term z, includes terms of the order z^2 and higher, they do not contribute to the expression for $\bar{\beta}^2$ up to the terms of the order z^3 and so, with the accuracy of the expression (115), $\bar{\beta}^2$ can be replaced by z^2. Then the threshold equation takes the form

$$\frac{E_{\gamma_b}^{th}}{E_{\gamma_b}^{Sth}}=\left(1-z^2\right)^{-b};\quad E_{\gamma_b}^{Sth}=\frac{m_e^2c^4}{E_\gamma} \quad (116)$$

where $E_{\gamma_b}^{th}$ is the modified value of the threshold and $E_{\gamma_b}^{Sth}$ is the standard value of the threshold. It is seen that the expression (116) for the threshold energy of the background photon differs from the standard one by the factor $(1-z^2)^{-b}$. The universal constant b is negative, both as it is expected from intuitive arguments and as it is found by fitting the cosmological model developed in the framework of the 'relativity with a preferred frame' to the observational data [27]. Thus, the threshold energy of the background photon decreases with the distance to the source (the redshift z).

Attenuation of gamma rays with the energy E_γ from the source at redshift z_s due to the pair production process is characterized by the optical depth $\tau_\gamma(E_\gamma, z_s)$. For z_s not too large one typically has $\tau_\gamma(E_0, z_s) < 1$ so that the Universe is optically thin along the line of sight of the source and if it happens that $\tau_\gamma(E_0, z_s) > 1$ the Universe becomes optically thick at some point along the line of sight. The value z_h such that $\tau_\gamma(E_0, z_s) = 1$ defines the γ-ray horizon for a given E_0, and sources beyond the horizon tend to become progressively invisible as z_s further increases. The optical depth is evaluated by

$$\tau_\gamma(E_\gamma, z_s) = \int_0^{l_s(z_s)} dl\ K_{\gamma\gamma_b}(E_\gamma, l(z)) \tag{117}$$

where $K_{\gamma\gamma_b}(E_\gamma, l(z))$ is the γ-ray absorption coefficient, which represents the probability per unit path length, l, that a γ-ray will be destroyed by the pair-production process. The absorption coefficient is calculated by convolving the spectral number density $n_b(E_{\gamma_b}, z)$ of background photons at a redshift z with the cross section of the pair production process $\sigma((E_\gamma, E_{\gamma_b}, \theta, z)$ (θ is the angle between the direction of propagation of both photons) for fixed values of E_{γ_b} and θ and next integrating over these variables [55], as follows

$$K_{\gamma\gamma_b}(E_\gamma, z) = \int_{-1}^{1} d(\cos\theta)\ \frac{1 - \cos\theta}{2} \int_{E_{\gamma_b}^{th}}^{\infty} dE_{\gamma_b}\ n_b(E_{\gamma_b}, z)\sigma((E_\gamma, E_{\gamma_b}, \theta, z) \tag{118}$$

Then the integral over distance l in (117) is represented as an integral over z to arrive at the expression for the optical depth in the form

$$\tau_\gamma(E_\gamma, z_s) = \int_0^{z_s} dz\ \frac{dl(z)}{dz}\ K_{\gamma\gamma_b}(E_\gamma, z) \tag{119}$$

The threshold energy of background photons $E_{\gamma_b}^{th}$ taking part in the expressions (118) and (119) is corrected according to (116) such that $E_{\gamma_b}^{th}$ decreases with the distance to the source (the redshift z). The cumulative outcome of this phenomenon may result in measurable variations in the expected attenuation of the gamma rays flux reducing the expected flux.

The preferred frame effects may influence the optical depth also via the cosmological part of the expression (119). In the Robertson-Walker metric (82) (or (85)), the distance element dl is defined as $dl = a(t)d\chi$ where $a(t)$ is the scale factor and χ is the radial distance element defined by (83). These quantities are calculated based on the GR equations (more specifically, Friedman equations) which leads to the expression (96) for the radial distance χ where the parameters are to be specified according to the cosmological model accepted. Commonly the quantity $\frac{dl(z)}{dz}$ is calculated within the standard 'concordance' ΛCDM cosmological model, where the expression (96) is specified to $\Omega_K = 0$, $\Omega_\Lambda = 1 - \Omega_M$ and x_1 given by (97), which yields

$$\frac{dl(z)}{dz} = \frac{1}{H_0} \frac{1}{(z+1)\sqrt{1 - \Omega_M + \Omega_M(1+z)^3}} \tag{120}$$

In the cosmology of the relativity with a preferred frame, $\Omega_\Lambda = 0$ and $\Omega_K = 1 - \Omega_M$ and, upon using these values in (96), one has for $\frac{dl(z)}{dz}$ the following

$$\frac{dl(z)}{dz} = \frac{1}{H_0}\frac{a(t)}{a(t_0)}\frac{1}{\sqrt{1-\Omega_M+\Omega_M(1+z)}} \tag{121}$$

where the quantity $\frac{a(t)}{a(t_0)}$ is to be calculated using several other equations as it is done (for the particular case $\Omega_M = 1$) in equations from (100) to (104). Similar calculations for the general case $\Omega_M \neq 1$ lead to the expression for $\frac{dl(z)}{dz}$ represented as series in z, as follows

$$\begin{aligned}\frac{dl(z)}{dz} = \frac{1}{H_0}\Bigg(1 + \left(-2 - b - \frac{\Omega_M}{2}\right)z + \left(3 + 3b + \frac{3b^2}{2} + \Omega_M + \frac{3\Omega_M^2}{8}\right)z^2 \\ + \left(-3 - 4b - \frac{5b^2}{2} - \Omega_M - \frac{3\Omega_M^2}{8}\right)z^3\Bigg)\end{aligned} \tag{122}$$

In the concordance model relation (120), the value $\Omega_M = 0.31.$ obtained from the observational data (see [27] for references), is used. In the present model, there is an interval of allowed values of Ω_M and the corresponding values of b, within which the results fit both the SNIa and BAO data [27]. The curvature K_c in the present model is not obligatory zero but the value of $\Omega_M = 1$ corresponding to the flat universe is within the interval of allowed values of Ω_M. Although Eqs. (120) and (122) defining dependence $\frac{dl(z)}{dz}$ on z in the concordance model and in the present model look completely different, the corresponding dependencies practically coincide as it is seen from **Figure 3**. Thus, the preferred frame effects influence $\tau_\gamma(E_\gamma, z_s)$ only via the threshold value $E^{th}_{\gamma_b}$ in (118), like in other Lorentz-violating theories (see, e.g., [56–58]).

5.3 Astrophysical tests for vacuum dispersion and vacuum birefringence

In the literature on Lorentz violation, as major features of the behavior of electromagnetic waves in vacuum in the presence of Lorentz violation, vacuum dispersion and vacuum birefringence are considered. Astrophysical tests for vacuum dispersion of light from astrophysical sources seek differences in the velocity of light at different wavelengths due to Lorentz violation which should result in observed arrival-time differences. For differences in the arrival times of different wavelengths to be interpreted as caused by differences in the light velocities, explosive or pulsed sources of radiation that produce light over a wide range of wavelengths in a short period, such as gamma-ray bursts, pulsars, or blazars, are to be used. All those are point sources, which have the disadvantage (to impose constraints on Lorentz violation) that a single line of sight is involved, which provides sensitivity to only a restricted portion of space for free coefficients of the Lorentz violating models.

The same is valid for the present theory leading to the dispersion relation (72). In the case of the waves propagating along the x-axis (aligned with the anisotropy vector k), when $q_y = q_z = 0$ and $q_x = q$, the two routes (74) become

$$\omega = \pm\frac{c}{1 \pm k}q \tag{123}$$

which corresponds to the waves propagating in the opposite directions. For a wave propagating to the observer from a cosmological source, with the x-axis directed from the observer to the source, the group velocity is

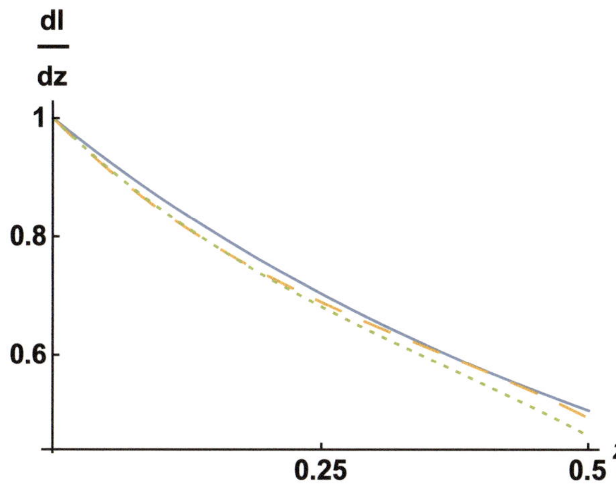

Figure 3.
The dependence of $\frac{dl(z)}{dz}$ (multiplied by H_0) on z for the concordance model with $\Omega_M = 0.31$ (solid) and for the cosmological model, based on 'relativity with a preferred frame' [27], with $\Omega_M = 1$, $b = -0.672$ (dashed) and $\Omega_M = 0.5$, $b = -0.495$ (dotted) where the values of the parameters Ω_M and b are chosen from those consistent with both the SNIa and BAO data (see [27]).

$$\frac{\partial \omega}{\partial q} = -\frac{c}{1-k} \tag{124}$$

It does not depend on q and so there is no place for vacuum dispersion.

Another test, that is commonly used for setting constraints on the parameters of the Lorentz-violating theories in electrodynamics, is the vacuum birefringence test. In birefringent scenarios, the two eigenmodes propagate at slightly different velocities. This implies that the superposition of the modes is altered as light propagates in free space. Since the two modes differ in polarization, the change in superposition causes a change in the net polarization of the radiation. However, it does not apply to the present theory leading to the dispersion relation (72). The two roots of the dispersion relation correspond to the waves propagating in different directions. Thus, no two eigenmodes are propagating in the same direction and so there is no possibility for vacuum birefringence. Thus, neither tests for vacuum dispersion nor tests for vacuum birefringence can impose restrictions, additional to those imposed by cosmological data, on the values of the only parameter of the theory b.

The vacuum birefringence and vacuum dispersion are widely discussed in the literature as astrophysical tests of Lorentz violation in the pure photon sector of the standard-model extension (e.g., [6, 38, 59–62]). Therefore it is of interest, in that context, to compare the Lorentz-violating terms, appearing in the Lagrangian due to the preferred frame effects in the present study, with those introduced as a formal SME extension. Extracted from the SME, the Lorentz-violating electrodynamics can be written in terms of the usual field strength F_{ik} defined by (57) and the potentials A^k, as follows

$$L = -\frac{1}{4}F_{ik}F^{ik} - \frac{1}{4}(k_F)_{nmik}F^{nm}F^{ik} + \frac{1}{2}(k_{AF})^n \varepsilon_{nmik}A^m F^{ik} \tag{125}$$

In what follows, we calculate the Lagrangian of the electrodynamics with a preferred frame and compare the Lorentz violating terms in that Lagrangian with those in (125). Calculating $L = -\frac{1}{4}F_{ik}F^{ik}$ using equations (58) and (59) yields

$$L = \lambda(k)^4 \left(\frac{1}{2} \left(\mathbf{E}^2 - \mathbf{H}^2 \right) + k \left(E_y H_z - E_z H_y \right) - k^2 \frac{1}{2} \left(E_y^2 + E_z^2 \right) \right) \tag{126}$$

It is seen that the form (126) is in a sense more general than (125) because of the Lorentz violating multiplier $\left(\lambda(k)^4\right)$. However, since the multiplier does not depend on the field variables and so does not influence the form of the field equations, it can be disregarded. Then the Lorentz-violating terms in (126) can be written based on (59) in terms of the field strength, as follows

$$L_{add} = k\left(F^{02}F^{12} + F^{13}F^{03}\right) - \frac{1}{2}k^2\left(F^{02}F^{02} + F^{03}F^{03}\right) \tag{127}$$

which fits the form (125) with the coefficients

$$(k_F)_{0212} = -4k, \ (k_F)_{0313} = -4k, \ (k_F)_{0202} = 2k^2, \ (k_F)_{0303} = 2k^2 \tag{128}$$

while other $(k_F)_{nmik}$ as well as all k_{AF} are zeros. The second term on the right-hand side of (125), not contributing to the Lagrangian of the present theory, could be disregarded from the beginning because it has theoretical difficulties associated with negative contributions to the energy [6, 38]. The Lagrangian defined by (127) (or (128)) provides an example of the Lorentz-violating SME (in a pure photon sector) which leads to equations of the electromagnetic wave propagation not exhibiting the vacuum birefringence and vacuum dispersion effects.

6. Discussion

The ‘relativity with a preferred frame’ incorporates the existence of the cosmological preferred frame into the framework of the theory while preserving fundamental principles of the SR: the principle of relativity and the principle of universality of the light propagation. The relativistic invariance is preserved in the sense, that the physical laws are covariant (their form does not change) under the group of transformations between inertial frames, and the relativistic symmetry is preserved (although modified) in the sense that there exists a combination, a counterpart of the interval of the standard relativity theory, which is invariant under the transformations. The existence of the modified symmetry provides an extension of the theory to general relativity such that the general covariance is also preserved. Thus, the ‘relativity with a preferred frame’ is a relativity theory, both in the special relativity and in the general relativity parts. Except for identifying the preferred frame with a comoving frame of cosmology, the theory does not include any assumptions. No approximations are involved besides approximating the universal function $k = F(\bar{\beta})$, defining dependence of the anisotropy parameter on the frame velocity relative to the preferred frame, by the expression $F(\bar{\beta}) = b\bar{\beta}$ valid up to the third order in $\bar{\beta}$. As the result, all the relations of the theory include only one universal parameter b.

The problem of defining allowed values of b is to be considered in the context of verification of the theory by observations since nothing in the theory itself imposes constraints on the values of b. Discussing the results of the application of the theory to natural phenomena, one can separate the conceptual and quantitative aspects. In the conceptual aspect, the cosmological models, developed using the modified general relativity, are of the most importance. First, it is related to the interpretation of the luminosity distance versus redshift relation deduced from the SNIa data, which has played a revolutionary role in the development of modern cosmology concepts. That relation, corresponding to the negative deceleration parameter, cannot be explained using cosmological matter-dominated models (Friedman-Robertson-Walker models) based on the standard general relativity. To explain the data, in modern cosmology, dark energy, a new type of energy with a peculiar equation of state corresponding to negative pressure, is introduced. In the cosmology of the 'relativity with a preferred frame', the luminosity distance versus redshift relation for the matter-dominated cosmological model contains corrections, such that the effective deceleration parameter can be negative. As the result, neither the acceleration of the universe expansion nor the dark energy providing the acceleration is needed. The consistency of the cosmological models, based on the 'relativity with a preferred frame', is supported by that, for any reasonable value of the parameter Ω_M, there exists a value of b such that the luminosity distance versus redshift relation fits with high accuracy the SNIa data.

In the applications of the theory to the BAO data, the conceptual and quantitative aspects go together. The BAO observations provide two different sets of data: BAO scales in transverse and line-of-sight directions. Measurements of the angular distribution of galaxies yield the quantity $D_M(z)$ which is the comoving angular diameter distance. Measurements of the redshift distribution of galaxies yield the value of the Hubble parameter $H(z)$. The fact that the two regions in the plane (Ω_M, b), within which the predictions of the present theory fit the D_M data and the $H(z)$ data, are overlapped, both provides a support for the theory and places quite tight constraints on the values of the parameters Ω_M and b since they should be confined within a quite narrow overlapping region. An additional (and quite strong) argument in favor of both consistency of the theory and estimates for the parameter b is that the line in the plane (Ω_M, b), on which the results of the present model fit the SNIa data, lies within that narrow region. Thus, the results fit well three different sets of observational data with the values of the theory parameter b confined within a quite narrow interval (approximately from $b = -0.4$ to $b = -0.8$).

Next, it might be expected that some constraints on allowed values of b could arise as the result of applying the theory to the cosmic rays data. In the propagation of the Ultra-High Energy Cosmic Rays from distant sources to Earth, the most remarkable effect is the attenuation due to pion photoproduction by UHECR protons which is characterized by the GZK threshold. Applying the 'relativity with a preferred frame' to the calculation of the energy threshold for the attenuation process results in the correction factor to the GZK limit. Although a comparison of that prediction of the theory with the data on the UHECR flux does not straightforwardly lead to constraints on the values of b, another issue, namely the data on the mass composition of UHECR, provides indirect confirmation of the theory. Those data, showing that the UHECR mass composition is dominated by protons only at energies around and below 10^{18} eV and then the fraction of protons is progressively decreasing up to energies of $10^{19.6}$ eV, contradict the previous consensus that UHECRs are mostly protons accelerated in the sources to $> 10^{20}$ eV. The prediction of the 'relativity with a preferred frame', that the GZK threshold energy decreases with the distance to the source of the particles (with the values of the parameter b defined by the cosmological data) allows to resolve, at least, partially, the contradiction between the view, that the primary UHECR flux is mostly protons accelerated to very high energies, and the observational data showing that the fraction of protons in the UHECR is decreasing towards higher energies. The explanation lies in that, because of decreasing the energy threshold with the distance to the source, the number of sources, contributing to the observed flux of protons at a given energy, should be progressively decreasing with the energy increasing.

Applying the modified particle dynamics to the pair-production process, which is responsible for attenuation of the gamma-rays flux, does not provide quantitative constraints on the values of the parameter b or indirect confirmations of the theory. At the same time, the results of applying the modified electromagnetic field dynamics to the behavior of electromagnetic waves in a vacuum maybe counted as a kind of indirect confirmation of the theory. The vacuum birefringence and vacuum dispersions are the features present in the popular Lorentz-violating theories (e.g., [6, 38, 59–62]) and the fact, that no indications of the existence of those phenomena are found in observations, imposes constraints on the values of numerous parameters of those theories. On the contrary, the electromagnetic field equations and based on them the electromagnetic wave equation of the present theory, although modified such that the Lorentz invariance is violated, does not predict such features as the vacuum birefringence and vacuum dispersion. Thus the absence of observational evidence for the existence of those phenomena may be considered as an argument in favor of the theory.

In general, the fact, that applying the theory containing only one universal parameter to several different phenomena does not lead to any contradictions, proves a consistency of its basic principles. The presence of only one parameter in the theory is a consequence of the fact that, as distinct from the popular Lorentz-violating theories, where Lorenz violation is introduced phenomenologically by adding Lorentz-violating terms to the Lorentz invariant relations, the 'relativity with a preferred frame' starts from the physically reasonable modification of the basic postulates of the SR. The generalized relativistic invariance, and so the Lorentz invariance violation, are ingrained in the theory at the most fundamental level being imbedded into the metric. It is also worth to emphasize that the conceptual basis of the theory has been developed without having in mind possible applications. It is aimed at designing the framework which would allow to incorporate the preferred frame into special relativity while retaining the relativity principle and the fundamental space-time symmetry. Nevertheless, the theory provides explanations of some observational data, that were regarded as puzzling after their discovery (like the SNIa luminosity distance-redshift relation indicating the acceleration of the universe and the absence of high energy protons in the UHECR flux). As the result, the concepts (among which dark energy is the most striking one), introduced to explain those puzzling features, become redundant. All the above justifies treating the 'relativity with a preferred frame' as an alternative to some currently accepted theories.

Author details

Georgy I. Burde
Alexandre Yersin Department of Solar Energy and Environmental Physics, Swiss Institute for Dryland Environmental and Energy Research, Jacob Blaustein Institutes for Desert Research, Ben-Gurion University of the Negev, Israel

*Address all correspondence to: georg@bgu.ac.il

References

[1] V. A. Kostelecky, S. Samuel, Spontaneous breaking of Lorentz symmetry in string theory, Phys. Rev. D. 1989;**39**:683

[2] V. A. Kostelecky, S. Samuel, Gravitational phenomenology in higher–dimensional theories and strings. Phys. Rev D. 1989;**40**:1886

[3] V. A. Kostelecky, Ed., Proceedings of the "Second Meeting on CPT and Lorentz Symmetry", Bloomington, Usa, 15-18 August 2001} (World Scientific, Singapore 2002)

[4] D. Oriti, Approaches to Quantum Gravity: Toward a New Understanding of Space, Time and Matter, (Cambridge University Press, Cambridge, UK 2009).

[5] D. Colladay, V. A. Kostelecky, CPT violation and the standard model, Phys. Rev. D. 1997;**55**:6760

[6] D. Colladay, V. A. Kostelecky, Lorentz-violating extension of the standard model, Phys. Rev. D; 1998;**58**:116002

[7] V. A. Kostelecky, Gravity, Lorentz violation, and the standard model. Phys. Rev. D. 2004;**69**:105009

[8] T. Jacobson, D. Mattingly, Gravity with a dynamical preferred frame, Phys. Rev. D. 2001;**64**:024028

[9] T. Jacobson, D. Mattingly, Einstein-aether waves, Phys. Rev. D. 2004;**70**: 024003

[10] T. Jacobson, Einstein-æther gravity: A status report, In: Proceedings of the conference "From Quantum to Emergent Gravity: Theory and Phenomenology", June 11-15 2007, SISSA; Trieste, Italy; v.2: PoS QG-Ph:020,2007, Sissa Medialab srl Partita IVA.

[11] J. Oost, S. Mukohyama, A. Wang, Constraints on Einstein-aether theory after GW170817, Phys. Rev. D. 2018;**97**: 124023

[12] T. Jacobson, S. Liberati, D. Mattingly, Threshold effects and Planck scale Lorentz violation: Combined constraints from high energy astrophysics, Phys. Rev. D. 2003;**67**:124011

[13] S. R. Coleman, S. L. Glashow, High-Energy Tests of Lorentz Invariance, Phys. Rev. D. 1999;**59**:116008

[14] S. Liberati, Tests of Lorentz invariance: A 2013 update, Classical quantum Gravity. 2013;**30**:133001

[15] R. Aloisio, P. Blasi, P. L. Ghia, A. F. Grillo, Probing the structure of space-time with cosmic rays, Phys. Rev. D. 2000;**62**:053010

[16] D. Mattingly, Modern tests of lorentz invariance, Living Rev. Rel. 2005;**8**:5

[17] S. T. Scully, F.W. Stecker, Lorentz invariance violation and the observed spectrum of ultrahigh energy cosmic rays, Astropart. Phys. 2009;**31**:220

[18] X. J. Bi, Z. Cao, Y. Li, Q. Yuan, Testing Lorentz invariance with ultra high energy cosmic ray spectrum, Phys. Rev. D. 2009;**79**:083015

[19] L. Maccione, A. M. Taylor, D. M. Mattingly, Planck-scale Lorentz violation constrained by Ultra-High-Energy Cosmic Rays JCAP04. 2009;**04**:022

[20] S. T. Scully, F. W. Stecker, Testing Lorentz invariance with neutrinos from ultrahigh energy cosmic ray interactions. Astropart. Phys. 2011;**34**: 575

[21] A. Saveliev, L. Maccione, G. Sigl, Lorentz invariance violation and chemical composition of ultrahigh-energy cosmic rays, JCAP03. 2011;**03**: 046

[22] F.W. Stecker, S. Scully, S. Liberati, D. Mattingly, Searching for traces of Planck-scale physics with high energy neutrinos. Phys. Rev. D. 2015;**91**:045009

[23] D. Boncioli, et al., Future prospects of testing Lorentz invariance with UHECRs, In: Proceedings of the 34th International Cosmic Ray Conference, 30 July- 6 August, 2015 The Hague, The Netherlands PoS(ICRC2015)521. Sissa Medialab srl Partita IVA.

[24] F.W. Stecker, Testing Lorentz symmetry using high energy astrophysics observations, Symmetry. 2017;**9**:201

[25] D. Boncioli for the Pierre Auger Collaboration, Probing Lorentz symmetry with the Pierre Auger Observatory, Proceedings of the 35th International Cosmic Ray Conference, 10-20 July, 2017 Bexco, Busan, Korea, PoS(ICRC2017)561. Sissa Medialab srl Partita IVA.

[26] R.G. Lang for the Pierre Auger Collaboration, Testing Lorentz Invariance Violation at the Pierre Auger Observatory, Proceedings of the 36th International Cosmic Ray Conference, July 24th - August 1st, 2019 Madison, WI, U.S.A. PoS(ICRC2019)327. Sissa Medialab srl Partita IVA.

[27] G.I. Burde, Cosmological models based on relativity with a privileged frame. Int. J. Mod. Phys. D. 2020;**29**: 2050038

[28] G.I. Burde, Particle dynamics and GZK limit in relativity with a preferred frame Astropart. Phys. 2021;**126**:102526

[29] G.I. Burde, Special Relativity with a Preferred Frame and the Relativity Principle. J. Mod. Phys. 2018;**9**:1591

[30] G.I. Burde, Special relativity kinematics with anisotropic propagation of light and correspondence principle. Found. Phys. 2016;**46**:1573

[31] H.P. Robertson, Postulate versus observation in the special theory of relativity. Rev. Mod. Phys. 1949;**21**:378

[32] R. Mansouri and S.U. Sexl, A test theory of special relativily: I. Simultaneity and slow clock synchronization, II. First order tests; III. Second order tests. Gen. Rel. Grav. 1977; **8**:497. 515, 809

[33] R. Anderson, I. Vetharaniam, G.E. Stedman, Conventionality of synchronisation, gauge dependence and test theories of relativity. Phys. Rep. 1998; **295**:93

[34] E. Minguzzi, On the conventionality of simultaneity. Found. Phys. Lett. 2002;**15**:153

[35] G.W. Bluman and S. Kumei, Symmetries and Differential Equations, Applied Mathematical Sciences, Vol. 81.} (Springer-Verlag, New York 1989).

[36] P.J. Olver, Applications of Lie Groups to Differential Equations (Graduate Texts in Mathematics: vol 107) (Springer, New York 1993).

[37] L.D. Landau, E.M. Lifshitz, The Classical Theory of Fields (Pergamon Press, Oxford 1971).

[38] Carroll, S. M., Field, G. B., and Jackiw, R., Phys. Rev. D. 1990;**41**:1231

[39] Kostelecky', V. A., and Mewes, M., Signals for Lorentz violation in electrodynamics, Phys. Rev. D, 2002;**66**: 056005

[40] Kostelecky', V. A., and Mewes, M., Electrodynamics with Lorentz-violating operators of arbitrary dimension., Phys. Rev. D, 2009;**80**:015020

[41] S. Weinberg: Gravitation and cosmology: principles and applications of the general theory of relativity. John Wiley \& Sons, Inc. New York London Sydney Toronto (1972).

[42] S. Weinberg: Cosmology, Oxford University Press, Oxford; 2008

[43] M. Betoule et al., Improved cosmological constraints from a joint analysis of the SDSS-II and SNLS supernova samples. A\&A. 2014;**568**: A22

[44] S. Alam et al., The clustering of galaxies in the completed SDSS-III Baryon Oscillation Spectroscopic Survey: cosmological analysis of the DR12. Mon. Not. R. Astron. Soc. 2017; **470**:2617

[45] R. Aloisio, Acceleration and propagation of ultra-high energy cosmic rays, Prog. Theor. Exp. Phys. 2017;**12**: 12A102

[46] R. Aloisio, P. Blasi, I. De Mitri, S. Petrera, (2018) Selected Topics in Cosmic Ray Physics. In: Aloisio R., Coccia E., Vissani F. (eds) Multiple Messengers and Challenges in Astroparticle Physics. (Springer International Publishing AG, Switzerland, 2018), pp. 1–96.

[47] L. A. Anchordoqui, Ultra-High-Energy Cosmic Rays Phys. Rept. 2019; **801**:1

[48] K. Greisen, End to the cosmic-ray spectrum? Phys. Rev. Lett. 1966;**16**:748

[49] G.T. Zatsepin, V.A. Kuzmin, Upper limit of the spectrum of cosmic rays, Pis'ma Zh. Ekps. Teor. Fiz. 4 (1966) 114 [JETP Lett. 4 (1966) 78].

[50] A. Aab et al., Pierre Auger Collaboration, Depth of maximum of air-shower profiles at the Pierre Auger Observatory II: Composition implications, Phys. Rev. D. 2014;**90**: 122006

[51] A. Aab et al., Pierre Auger Collaboration, Evidence for a mixed mass composition at the "ankle" in the cosmic-ray spectrum, Phys. Lett. B. 2016;**762**:288

[52] R. Abbasi et al., Study of ultra-high energy cosmic ray composition using telescope Array's Middle Drum detector and surface array in hybrid mode, Astropart. Phys. 2015;**64**:49

[53] R. Aloisio, V. Berezinsky, S. Grigorieva, Analytic calculations of the spectra of ultra-high energy cosmic ray nuclei. I. The case of CMB radiation, Astropart. Phys. 2013;**41**:73

[54] R. Aloisio, V. Berezinsky, S. Grigorieva, Analytic calculations of the spectra of ultra-high energy cosmic ray nuclei. II. The general case of background radiation, Astropart. Phys. 2013;**41**:94

[55] De Angelis, A.; Galanti, G.; Roncadelli, M. Transparency of the Universe to gamma rays. Mon. Not. Roy. Astron. Soc. 2013;**432**:3245

[56] Lang, R.G., Marténez-Huerta, H., de Souza, V. 2018, Limits on the Lorentz Invariance Violation from UHECR astrophysics. Astrophys. J., 2018;**853**:23

[57] Lang, R.G., Marténez-Huerta, H., de Souza, V. 2019, Improved limits on Lorentz invariance violation from astrophysical gamma-ray sources. Phys. Rev. D. 2019;**99**:043015

[58] Marténez-Huerta, H., Lang, R.G., de Souza, V. 2020, Lorentz Invariance Violation Tests in Astroparticle Physics Symmetry, 2020;**12**:1232

[59] Kostelecky', V. A., and Mewes, M., Cosmological Constraints on Lorentz Violation in Electrodynamics. 2001;**87**: 251304

[60] Kostelecky', V. A., and Mewes, M., Sensitive Polarimetric Search for Relativity Violations in Gamma-Ray Bursts, 2006, Phys. Rev. Lett. 2006;**97**: 140401

[61] Kostelecky', V. A., and Mewes, M., Lorentz-Violating Electrodynamics and

the Cosmic Microwave Background, 2007, Phys. Rev. Lett. 2007;**99**:011601

[62] Kostelecky', V. A., and Mewes, M., Astrophysical Tests of Lorentz and CPT violation with Photons, 2008, Astrophys. J. 2008;**689**:L1

Chapter 8

Dark Energy as an Information Field and Attribute of the Universe

Alexander Bogomolov

Abstract

The hypothesis that information is the same attribute of the Universe as space, time and matter, which appeared after the Big Bang and its carrier is the dark energy, is proposed. We assume that along with the development of matter and space–time continuum the development of information took place, i.e. its accumulation and complication of forms of its realization in the Universe, and the formation of information field. The proposed concept is close to or coincides with the bio-centrism concept. They substantiate the connection between information and dark energy from philosophical and physical positions. Including properties of "black holes" of the Universe. The rate of increase of knowledge and volumes of created information in the process of human evolution is very similar to the rate of increase of "mass" of dark energy of the Universe. They concluded that structured information in the information field, which forms the noosphere or thinking shell, plays a decisive role both in life of an individual and humanity. Comparison of quantum properties of information field of the Universe (noosphere) and dark energy may lead to new discoveries of the essence of life and the Universe. Studying and mastering of noosphere will mean the transition of humanity to a cosmic stage of its development.

Keywords: big bang, human birth, dark energy, information field, noosphere, black holes

1. Introduction

We live in a universe that is expanding and developing. But once upon a time, the Universe, as most scientists and religious thinkers believe, did not exist, and it arose from something or was created. There are many scientific theories of the origin of the universe, but most scientists have abandoned the idea of an infinite and timeless universe since the 20th century. The most developed, widespread and confirmed by indirect evidence recently got because of new astronomical and cosmological discoveries is the Big Bang theory. According to this theory, the Universe was born because of the Big Bang of some singularity (point) having structure and content about which we know nothing, except that the laws of physics known to us did not operate in it. We will probably never know the cause of the explosion, which occurred about 13.8 billion years ago. After the Big Bang came time, space, which expanded, and matter as elementary particles that later formed various elements (matter). The scattered tiny elements gave rise to the world we know today. After the Big Bang, the universe expanded and is still expanding today (**Figure 1**).

Figure 1.
The birth of the universe. Source: Yandex.ru.

The main disadvantage of the Big Bang theory is the inability to explain the cause of the explosion and the mechanisms of the origin of space–time-matter [1]. Also, the Big Bang theory has difficulties when trying to describe the behavior of the Universe from the position of quantum mechanics. I believe that since the Big Bang emerged such attributes of the universe as space, time, matter. In modern science, several scientists believe that information may also be one attribute of the Universe [2–5]. It is logical to assume that along with the development of matter and space–time continuum, there was the development of information, i.e. its accumulation and complication of forms of its realization.

The modern level of scientific cognition increasingly allows us to connect the progressive development of matter with the processes of reflection and with the accumulation of structural information. The accumulation of information in the process of life not only in an individual person but also in the evolutionary process of the development of society, the transformation of information into knowledge, is characterized in the last 100 years by an explosive character. By 2020 we estimate the accumulated

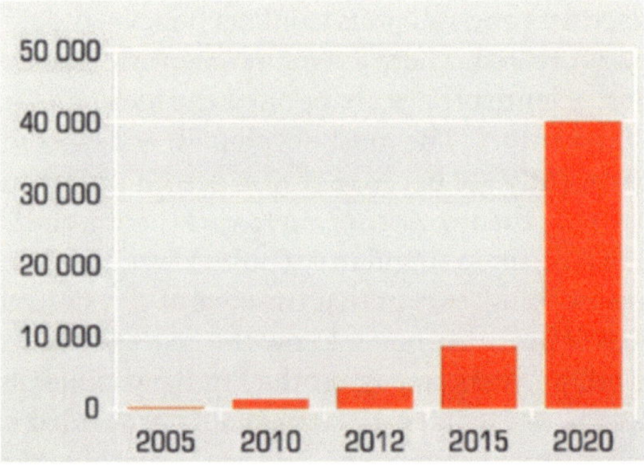

Figure 2.
Explosive increase in information (source IDC).

volume of information at about 40 zettabytes, which is 57 times more than the number of grains of sand on all the beaches of the planet (**Figure 2**).

The growth rate of knowledge and volumes of information created in the process of the evolution of mankind is very similar to the growth rate of the "mass" of the dark energy of the Universe. Its special properties allow it to claim the role of a carrier of information field of the Universe.

2. Dark energy—the main riddle of the fundamental physics of the XXI century

2.1 Religion about the creation of the world

The theories of the origin of the universe in various religious movements agree that it was created as a result of a conscious and spiritual action by God or the Creator. The major difference between scientific and religious theories of the origin of the universe is primarily in terminology. Scientific theories use the term "origin" instead of "creation," the term "Nature" instead of "Creator," etc. One can find many similarities between scientific and religious theories of the origin of the universe.

Religious teachings, in fact, have not strayed that far from the model of the creation of the world by the Big Bang. For example, the Bible says that the Word (information) was at the beginning of creation. The question of the reason for the appearance (creation) of the World has concerned thinkers since the fourth century A.D. 1600 years ago, the theologian Augustine the Blessed tried to understand the nature of God before the creation of the universe. He concluded that time was part of God's creation, and there simply was no "before" the creation of the world. The belief in God's creation of the world out of nothing is the basic doctrine of creation in traditional Christian theology.

2.2 String theory

The theory explaining the origin of the universe because of the Big Bang is not the only one. For example, string theory assumes a cyclical birth and death of the universe2. According to string theory, they do not build the entire world of particles, as we believe it today, but of infinitely thin objects with the ability to make vibrations, which allows us to draw an analogy with strings. With the help of complex mathematical apparatus it is possible to connect these vibrations with energy and hence with mass any particle arises because of one or another type of quantum string vibrations.

Like any unproven theory, string theory has several difficulties, which suggests that it requires enormous refinement.

2.3 Dark matter and dark energy

Scientists today believe that matter (that we see around us and in the Universe) makes up only less than 5% of the known Universe. The rest is a dark matter at 25% and about 70% is dark energy.

Dark matter is akin to ordinary matter in the sense that it can clump together (say, the size of a galaxy or galaxy cluster) and takes part in gravitational interactions in the same way as ordinary matter. Most likely, it comprises new particles, not yet discovered in terrestrial conditions. This itself testifies that there is a new, not yet discovered law of conservation in nature that forbids these particles to disintegrate. In our Galaxy, near the Sun, the mass of dark matter is approximately equal to the mass of ordinary matter. These particles must accumulate in the centers of the planets and in the center of the Sun (the matter is practically transparent for them, and they can fall inside the Earth or the Sun). There they annihilate with each other, and other particles, including neutrinos, are produced.

Dark energy is a much stranger substance than dark matter. The mystery of dark energy originated much earlier than the now common term. The starting point of this story was astronomers' estimates of the masses of various galaxies. The estimated mass of the galaxies was about 10 times the total mass of all their stars. This means that some unknown substance handles this mass [6–9]. This substance is called "dark energy". It is the one that dominates the Universe.

They do not collect dark energy in clumps, but is uniformly "spilled" in the Universe. In galaxies and clusters of galaxies there is just as much of it as outside of them. The most unusual thing is that dark energy generates antigravity. This distinguishes it dramatically from ordinary forms of matter. The accumulation and increase of dark energy leads to the expansion of the universe.

2.4 Dark energy and the expansion of the universe

References astronomical observations show that today (and in the not too distant past) the Universe is expanding with acceleration: the rate of expansion increases with time [10]. In this sense we can talk about antigravity: the usual gravitational attraction would slow the expansion of galaxies, but in our Universe, it turns out that it is the opposite. Dark energy handles the expansion of the Universe with modern astronomical methods it is possible not only to measure the current rate of expansion of the Universe, but also to determine how it has changed. What can be the reason for the increase of an antigravity mass of dark energy, what processes in the Universe also occur at an increasing rate? One of such processes is the process of generation, accumulation and transformation of information in the Universe, in particular on the Earth. What is information?

2.5 Information and the information field

"Information is information, not matter and not energy" -N. Wiener.

Despite its widespread occurrence, the concept of information remains one of the most debatable in science, and the term may have different meanings in different branches of human activity. The modern level of scientific cognition increasingly allows us to link the progressive development of matter with the processes of reflection and with the accumulation of structural information (knowledge). Information, as a measure of ordering of material structures and their interaction, is an objective participant at all stages of organization of matter. Information participated in the processes of self-organization of matter, contributing to the emergence of life. The most important epistemological problem is understanding the place and role of information in the universe and human life.

One of the most common concepts in physics is the concept of a field. It also used this concept in many other fields of knowledge: cosmogony, astrophysics, quantum physics, biology, etc. The concept of a field also applies to information. The mathematical theory of the information field (IFT) is based on a serious mathematical apparatus [11].

The IFT theory uses computational methods developed for quantum field theory and statistical field theory, allows mathematical derivation of optimal visualization algorithms, data analysis methods and even computer simulation schemes.

Application of IFT algorithms to astronomical datasets provides high accuracy images of the Universe and facilitates the search for weak statistical signals from the Big Bang.

2.6 Noosphere—the information field of the universe

They commonly referred the information field of the Universe to as the noosphere. B. Vernadsky introduced the concept of the noosphere as the highest form of development of the biosphere [12]. The central theme of the doctrine of the noosphere is the unity of the biosphere and mankind. Vernadsky in his works reveals the roots of this unity, the importance of the biosphere's organization in the development of mankind. This allows us to understand the place and role of historical development of humanity in the biosphere's evolution, the regularities of its transition to the noosphere.

One of the key ideas underlying Vernadsky's theory of the noosphere is that man is not a creature living separately according to its own laws, he exists within nature and is part of it. This unity is primarily because of the functional inseparability of the environment and man, which Vernadsky tried to show. Humanity itself is a natural phenomenon, and it is natural that the influence of the biosphere affects not only the environment of life but also its thinking sphere.

But it is not only nature that affects man; there is also a feedback. And it is not superficial, reflecting the physical influence of man on the environment, it is much deeper. This is proved because recently the planetary geological forces have noticeably intensified. " ... we are seeing more and more vividly the geological forces around us in action. This has coincided, hardly coincidentally, with the penetration into the scientific consciousness of the belief about the geological significance of *Homo sapiens*, with the identification of a new state of the biosphere - the noosphere - and is one form of its expression. It is connected, of course, first with the clarification of natural scientific work and thought within the biosphere, where living matter plays a major role".

Developing V. Vernadsky's ideas, P. Teilhard de Chardin [13] called the noosphere a certain "shell of thought" over the Earth. He imagined mind as a flame in which the globe warms up and which gradually covers the planet, forming its new cover: "The Earth is not only covered by myriads of thought grains, but is enveloped by a single thinking shell that forms functionally one vast thought grains on a cosmic scale. The multitude of individual thoughts is grouped and amplified in the act of one unanimous thought.

The "thought shell" over the Earth, or the Noosphere, includes a multitude of individual thought-images (the author's definition) of individual living beings of the Earth, including humans (**Figure 3**).

One can consider that the biofield of an individual human being is his individual noosphere. **Figure 4** shows a symbolic image of man and his noosphere.

Human noosphere is a structural unit of noosphere of the Earth, and the latter in its turn is a structural unit of noosphere of the Cosmos (**Figure 5**).

Such a hierarchical network structure of the Universe noosphere testifies to the universal connection of events occurring in it. Astrophysicist V.A. Ambartsumyan once said: "I sneeze and the whole cosmos will shake" [14]. By the way, the modern quantum theory (Einstein-Podolsky-Rosen paradox) [15] does not deny it. The EPR paradox speaks of the presence of a single wave function of correlated (bound) quantum particles, i.e., of the universal connection of phenomena in nature.

The eminent scientist and naturalist N. Tesla believed that he received his scientific revelations by drawing information from the single information field of the Earth (noosphere) [16].

Figure 3.
Thought-images of a person. Source: Yandex.ru.

Figure 4.
Thought-images of a man form a noosphere around him. Source: Yandex.ru.

Being convinced that the Universe is alive and humans are to a certain extent "automatons" behaving according to the Creator's plans, Tesla proposed an original theory of memory. He believed that the human brain does not possess the ability

Figure 5.
Noosphere of the earth. Source: Yandex.ru.

to remember in the sense that it is commonly believed (biochemically, or rather biophysically), and that memory is merely the ability of the human brain to retrieve this or that information (thought-image) from the noosphere.

Another eminent scientist, Albert Einstein, also believed that our universe is rational. "I want to know how God created this world. I want to know his thoughts, the rest is particular," Einstein said [17].

2.7 Biocentrism

We should mention the doctrine of biocentrism [18]. Proponents of biocentrism argue that current theories of the physical world do not work. While physics is foundational to the study of the universe, chemistry is foundational to the study of life, scientists need to place biology before the other sciences to get a theory of everything. The theory of bio-centrism is based on the idea of the primacy of cosmic consciousness, cites in its proof the achievements of quantum physics.

Something specially created the world for the existence of life, not only on a microscopic subatomic level but also on the scale of the entire universe. Wherever life appears, the world grows around it.

Scientists have already found many indications that all matter in the universe, from atoms to stars, seems to have been tailor-made just for us. Other researchers speak of the principle of intelligent design and believe that the cosmos is such a fitting place for us for a reason. True, the second concept is a veritable Pandora's Box, for it is a fertile field for all kinds of biblical interpretations and references to other topics, which in this case are completely unimportant or worse. Whatever these new discoveries are called, they have already sparked some serious discussions both within and outside the professional astrophysics community.

In the late 1960s, it had already become clear that if the Big Bang had been a millionth stronger, cosmic matter would have been blown around too quickly, and stars and planets would not have had time to form. As a result, we would not have existed. Even more incredible coincidence seems that all four basic interactions existing in the Universe and all constants are perfect for the flow of interatomic interactions, the existence of atoms and chemical elements, planets, liquid water and life. It is enough to change any of these constants slightly and we would not exist.

Such life-affirming physical values are woven into the fabric of reality like cotton and linen fibers into good currency. The best known of these constants is probably the gravitational constant, but the fine structure constant plays an equally key role in the existence of life. If this quantity, called "alpha, "were only 1.1 times or more above its present value, nuclear fusion could not take place in stars. The fine structure constant is causing so much scrutiny because the Big Bang formed only hydrogen and helium - and virtually nothing else. Oxygen and carbon are necessary for the existence of life (suffice it to say that oxygen is necessary for the appearance of water), but there is no shortage of oxygen precisely because it is formed in stars as one product of nuclear fusion.

The concept of biocentrism is based on the following principles.

- Perception of reality is a process in which our consciousness is directly involved our consciousness.
- Our external and internal sensations are inextricably linked. They cannot be cannot be separated into two sides of the same coin.
- The behavior of elementary particles - in fact any particle or object - is inextricably linked to an observer. In the absence of a conscious an observer, all elements of reality exist in an indeterminate state and are probabilistic waves.
- Without consciousness, "matter" is in an indeterminate probabilistic state. If the Universe existed before consciousness, it existed only in a probabilistic state.
- The entire organization of the Universe can only be explained from a biocentric perspective. The universe is finely tuned to support life, and it makes perfect sense that life creates the universe, not the other way around. The universe is simply a completely inconsistent space–time representation of itself.

2.8 The noosphere from the physical point of view

The manifestation of noosphere as information field, connected with the events in the nature and socio-economic sphere, allows us to speak about their universal connection. Nevertheless, the key question is still the nature and essence of noosphere.

The attempt to find an explanation of the noosphere as an information field from the physical point of view or from the point of view of a scientific materialist is of great interest. For example, the existence of the noosphere agrees also with the notion of black holes in modern physics. Nothing, even the smallest particle of light, can avoid the fate of being absorbed into a black hole if it is nearby. The existence of black holes is proven and we observe them in the Universe. But what happens to an object that is caught inside a black hole, physicists do not know and only speculate.

Previously, theoretical physicist Stephen Hawking suggested that black holes can indeed destroy the entire essence of an object, leaving behind only a tiny quantum trace (electric charge or spin). But then where does the information go? In the late 1990s, Hawking rejected the idea that black holes extinguish information. Instead, the scientist suggested that information may indeed still exist, but in a very different form [19].

DOI: http://dx.doi.org/10.5772/intechopen.99044

The expansion of the Universe testifies to a constant increase in the "mass" of dark energy since the beginning of the Big Bang, and this increase occurs at an increasing rate. Since the Big Bang the structure of the Universe, i.e. the increase of its information or non-entropy, has also been increasing at an increasing rate. In April 1982 (!) Academician M.A. Markov reported at the Presidium of the USSR Academy of Sciences: " ... the information field of the Universe is layered and structurally resembles a "matreshka", and each layer is linked hierarchically with higher layers, up to the Absolute, and besides an information bank it is also a regulator of the beginning in the fate of people and humanity" [20]. And if the information field of the Universe really exists, there is no better candidate for its material embodiment than dark energy.

2.9 The fractal principle of the universe

The development of the Universe and complication of its structure can be linked with increase of its non-entropy (information) but its entropy is increasing which eventually will lead to its thermal death. These two processes resemble by analogy the birth, development and death of a human, which is also a kind of the Universe.

It manifested fractal principle of the Universe construction because structure and form of some objects are similar or repeated in other objects, including those of quite original nature. For example, some scientists find similarity in the structure of galactic networks and neural networks of human brain. Fractal principle also takes place during realization of evolutionary processes of development of absolutely different observable systems. It is possible to find a deep analogy between birth, life and death of a human, and between birth, development and death of the Big Universe [21].

2.10 Similarity between the development of man and the universe

Human birth also resembles the Big Bang of some singularity - the size of an egg is about 0.1–0.15 microns. Then man grows, his mass increases, and is complexity as a system grows, with accumulation of his non-entropy (information). Eventually man is destroyed (dies), his entropy reaches a maximum, and his non-entropy reaches a minimum or zero.

The growth of non-entropy occurs in an open system because of inflow of energy and information from outside. Similar processes take place, apparently, in the cosmic Universe. Accumulation of information and knowledge on Earth, as well as in other intelligent worlds of the Universe, takes place with increasing pace, as well as the growth of dark energy mass, which gives reason to hypothesize about their strong interconnection or identity.

3. Conclusion

If dark energy is the information field of the universe.

Most scientists are of the opinion that, besides matter, there is in our world some as yet unknown to science substance, which creates antigravity, almost an order of magnitude greater than the gravity of the material component of the Universe. Before this discovery, astronomy and astrophysics were mainly concerned with the study of the material Universe, comprising atomic matter and radiation. After it will be finally revealed that it is not this part of the Universe that dominates, but a substance of immaterial nature, as yet unknown to science, it will be necessary to make a major revision of current cosmological ideas about our world, which will lead to a change in the cosmological paradigm, established based on discoveries of the first quarter of the twentieth century.

Someone practically accepted that humanity is at the stage of transition from post-industrial to information society. Information society is such a stage of society development, when the use of information and communication technologies (ICT) has a decisive influence on the main social institutions and spheres of life Information society is a special society, unknown to history. It is difficult or impossible to say what humanity will look like in, say, 100 or 200 years. However, we can probably agree with K.E. Tsiolkovsky that humanity humanity will pass to the "ray form" of its existence and will be spread in such form throughout the Universe [22].

If to suppose that already now informational essence of each person separately, humanity and informational essence of other inhabited worlds are already in the informational field of the Universe - noosphere, then "mastering" of noosphere will mean transition of humanity to the space stage of its development. The transition to the "ray form" or noosphere will occur at some stage of human development. By the way, this may explain why we have not yet met our brothers in mind. Because of the exponential growth of knowledge, intelligent civilizations do not reach the stage of building interstellar ships and interstellar travel. Even earlier, they master knowledge from the cosmic noosphere and go to the ray form, becoming practical gods. The same is waiting for humanity, unless it destroys itself earlier. In the radiant form it is possible to control matter, transforming it into various forms. Perhaps, the results by intelligent entities are displayed in the phenomena that we call UFOs. Otherwise, from the point of view of today's science, it is impossible to explain the appearance, appearance and behavior of UFOs.

Philosophical understanding of the noosphere concept leads to a conclusion formulated by Georg Hegel "Everything real is reasonable, everything reasonable is real". The hypothesis about a close connection between dark energy and the informational field of the Universe, if it finds confirmation, will lead to revolutionary changes in the basic provisions of science, the appearance of new fabulous technologies and radical revision of the tasks and lifestyle of society, putting not material but spiritual values at the top of the list.

Author details

Alexander Bogomolov
Financial Institute under the Government of the Russian Federation,
Moscow, Russia

*Address all correspondence to: aibogomolov@fa.ru

References

[1] Hawking S. et al. Theory of Everything. Origins and destiny of the Universe. Stephen Hawking; [translated from English by I. Ivanov; edited by G. Burba]. SPb: Amphora. Amphora TID, 2009. p. 148.

[2] Zummermann R.E. Matter and information as attributes of substance. The European Physical Journal Special Topics 226(2), 2017. DOI:10.1140/epjst/e2016-60365-0, pp. 177-180

[3] W. Hofkirchner. Introduction: Information from physics to social science. The European Physical Journal Special Topics, 226(2), 2017. pp. 157-159. DOI: 10.1140/epjst/e2016-60373-6

[4] Khanzhin A.G., Kojocaru A.A. Revision of the concept of information. STI. Ser. 2. No. 6, 2008. pp. 1-9.

[5] Belonogov A.I., Gilyarevsky R.S., Khoroshilov A.A. On the Nature of Information, STI. Ser. 2. No. 1, 2009. pp. 1-9.

[6] Astrophysical Journal, 1985, v. 299, p. 102.

[7] Physical Rev. Letters (US), 1986, v. 59, N 3, p. 263-265

[8] Astronomy and Astrophysics, v. 236, p. 99-106. 1990

[9] Overbye, Dennis (February 25, 2019). "Have Dark Forces Been Messing With the Cosmos?- Axions? Phantom energy? Astrophysicists scramble to patch a hole in the universe, rewriting cosmic history in the process." The New York Times. Retrieved February 26, 2019.

[10] Oyvind Gron (December 2018). The Discovery of the Expansion of the Universe. Galaxies 6(4): 132. DOI: 10.3390/galaxies6040132.

[11] Enßlin, Torsten A. (2019). Information theory for fields. Annalen der Physik. 531 (3). doi: 10.1002/andp.201800127

[12] Vernadsky, V. I. Scientific thought as a planetary phenomenon. Moscow: Nauka, 1991. 271

[13] Noosphere of Teilhard de Chardin and Vernadsky.

[14] Gordina L.S. On the modern concept of "Noosphere". Available at: http://noospherecity.com/o-sovremennom-ponyatii-noosfera

[15] Crisis in fundamental physics. Is there a way out? Available at: http://prometheus.al.ru/phisik/95kriz.htm

[16] Cheney, Margaret 2001 [1981]. Tesla: Man Out of Time. Simon and Schuster. ISBN 978-0-7432-1536-7.

[17] Gilmore M. So what did Einstein believe in Einstein on Religion. M. Alpina non-fiction, 2010. p. 133.

[18] Derr, Patrick George; Edward M. McNamara (2003). Case studies in Environmental ethics. Rowman and Littlefield. p. 21. ISBN 978-0-7425-3137-6.

[19] S. Hawking, The Edge of the Universe, Nature, 1985, No. 4, p. 21.

[20] The energy-information field of the Universe. Available at: http://to-be-free.ru/eipvs.

[21] Bogomolov A.I. Noosphere and Dark Energy. Chronoeconomics. 2017. No 6 (8). pp. 7-11

[22] K.E. Tsiolkovsky. Radiant humanity. Part 1. Available at: https://michael101063.livejournal.com/5174.html